How Large
Is God?

Voices of Scientists and Theologians

How Large Is God?

Voices of Scientists and Theologians

 Edited by John Marks Templeton

TEMPLETON FOUNDATION PRESS

PHILADELPHIA & LONDON

Templeton Foundation Press
Two Radnor Corporate Center, Suite 320
100 Matsonford Road
Radnor, PA 19087

Printed in the United States of America

Library of Congress Number: 97-90755

ISBN 1-890151-01-7

Contents

How Large
Is God?

Voices of Scientists and Theologians

Introduction

JOHN MARKS TEMPLETON

This book, I hope, will help bring more people to the realization that we know very little—probably less than 1 percent of what can be discovered—about God and fundamental spiritual principles. In recent years, scientific research has revealed that the universe is staggering in its immensity and intricacy, and some scientists are now suggesting that a much larger God than we previously had imagined may be its source.

Yet, as a member of the board of trustees of Princeton Theological Seminary for thirty-nine years, I gradually realized that most people—even highly trained theologians—seem to have various, restricted views of who God is and what his purposes are in creating this amazing universe. Just in my lifetime, our knowledge in science and medicine has grown a hundredfold. More than half the scientists who ever lived are alive today. More than half the discoveries in the natural sciences have been made just in this century. More than $1 billion every day worldwide is now spent on scientific research.

Yet by comparison, almost nothing is spent on spiritual research, despite the enormous benefits that might come from even one-tenth that amount invested in research for spiritual information about meaning and purpose and the invisible realities. For this reason, twenty-five years ago the Templeton Prize for Progress in Religion (a cash prize larger than the Nobel Prize) began to be awarded annually to emphasize the benefits of spiritual progress. Eight of the recipients have been scientists or scholars interested in increasing spiritual information through science.

More recently, my colleagues and I began many new projects to further the goal of increasing spiritual information and research. In 1987, we established the John Templeton Foundation, the central aim of which is for humanity to gain the benefits of new spiritual information through science. The crucial ingredient in this research program,

3

to bring scientists and theologians and others to work together, is called Humility Theology. A description of this initiative follows and serves as an introduction to these nine essays by some of the scientists and theologians who have been our advisors in these early years of the development of the foundation's programs.

Humility Theology

At the time of the organization of the first Templeton Foundation in 1987, we proposed the idea of a Theology of Humility. This concept came about through the recognition that the sciences, in these past few decades, have opened our vision to realities invisible and intangible and a universe that is so vast and complex that our view of ourselves and of our Creator would seem to need a searching reassessment. The Theology of Humility, as formulated for the Bylaws of the Foundation,[1] emphasized that:

1. It is centered in an infinite God
2. It encourages creativity and progress.
3. It recognizes diversity and constant change as hallmarks of our universe.
4. It encourages research into spiritual subjects such as love, prayer, and thanksgiving in what might be seen as a kind of "experimental theology."

Scholarship seems so often to be associated with pride. Knowing more than someone else has seemed for many intelligent people to be almost more important than the acquisition of knowledge itself. This tendency of the human ego has been especially destructive to progress in our information about God. The founding prophets of the great religions often took an exclusive view of their knowledge of God. They assumed that there was little new to be learned and that their audience could comprehend and remember only limited spiritual information. Research has often been backward-looking, focusing on the ancient foundations instead of the future.

In an interview with the publication *Second Opinion* in July 1993, I was asked why none of the great religions seems ready to experiment with openness. I responded:

> The main restraining influence has been and is personal ego—the concept that we are the center. For countless ages various people thought that the earth was flat, because it looks flat. For countless ages various people have thought that the sun revolved around the earth, because it looks that way. For countless ages people have thought that their god was the only true God. The Jews were not the only ones to think they were the chosen people. And the human ego has in effect said that God is understandable. Human ego has led most religions—I'm talking about forgotten religions—to say that they had the whole truth, they knew their mysteries. Now astronomy has defeated human ego—we no longer think we are very important in a hundred billion galaxies. I would like to see that happen in our knowledge of God. I don't think we know much more about God now than we knew about the hundred billion galaxies 2,000 years ago.[2]

Examination of the Bylaws

The first emphasis in the Bylaws statement of the John Templeton Foundation reads:

Centered in an Infinite God

The Theology of Humility means we know so little and need to learn so much and to devote resources to research. The Theology of Humility is not man-centered but God-centered.

It proposes that the infinite God may not be describable adequately in human words and concepts and may not be limited by human rationality. Perhaps God is not limited by our five senses and our perceptions of three dimensions in space and one dimension in time. Perhaps there was no absolute beginning and there will be no absolute end, but only everlasting change and variety in the unlimited purposes, freedom and creativity of God.

Maybe God is all of time and space—and much more. The appearance of humankind on this planet may be said to have heralded the coming of a new quality encircling the earth, the sphere of the intellect. Then as we have used our intellects to investigate this mysterious universe, accumulating knowledge at an ever-increasing rate, there has come a growing awareness that material things are not what they seem; that maybe thoughts are more real and lasting than matter and energy.

Perhaps this heralds a new quality, the sphere of the spirit. God may be creating not only the infinitely large but also the infinitely small; not only the outward but also the inward; not only the tangible but also the intangible. Thoughts, mind, soul, wisdom, love, originality, inspiration and enthusiasm may be little manifestations of a Creator who is omniscient, omnipotent, eternal and infinite. The things that we see, hear, and touch may be only appearances. They may be only manifestations of underlying forces including spiritual forces which may persist throughout all the transience of physical existence. Perhaps the spiritual world, and the benevolent Creator whom it reflects, may be the only reality.

Presumably the sphere of the spirit may enclose not only this planet but the entire universe and so God is all of Nature, and is inseparable from it. Perhaps it is mankind's own ego which leads us to think that we are at the center of a vast universe of being which subsists in an eternal and infinite reality which some call God. Maybe all of nature is only a transient wave on the ocean of all that God eternally is. Maybe time, space and energy provide no limit to the Being which is God. Likewise the fundamental parameters of the universe—the speed of light, the force of gravitation, the weak and strong nuclear forces and electromagnetism—would seem to pose no limits to the Being which is God.

These ideas with their focus on the infinity and magnificence of God, and of his inseparability from all that has been made, may be illustrated by the following tentative propositions for future research:

1. Maybe nothing can be separate from God.
2. Maybe the great religions which descended from Abraham have unconsciously overemphasized the visible and tangible aspects of fundamental reality.
3. Maybe unconsciously these religions have overemphasized God as distant and separate.
4. Maybe it is egotistical to think that any human ever understood even 1 percent of God.
5. Maybe humanity is like a wave on the ocean, and God is like the ocean.
6. Maybe it is self-centered for us to think that nowhere in the universe is there other life capable of thought. Even if only one star in a million has a planet in the same developmental stage as Earth, then there may be a hundred thousand other similar planets in our own Milky Way, not to mention the other hundred billion galaxies.

The second part of the description of Humility Theology as elaborated in the Bylaws:

Creativity and Progress

The Theology of Humility encourages change and progress. Science is revealing to us an exciting world in dynamic flux whose mechanisms are evermore baffling and staggering in their beauty and complexity. Scientists are learning to live and work with quantum uncertainty and complementarity, major discontinuities in evolution and baffling complexity in cellular differentiation. Yet scientists have turned these and other discoveries into opportunities, and many scientists have expressed a new openness to philosophical and religious questions about life and the universe.

While science has generally responded favorably to change, the long history of religion is filled with the failures of thousands of religions. Perhaps these religions disappeared partly because their conceptions of God were too small. Theology of Humility proposes that maybe God is now providing new revelations in ways which go beyond any religion, to those who welcome the originality of the Creation and its continual surprises. For example, some theologians and scientists see an integration of the new discoveries of science with many religions' traditions—a new "theology of science."

Perhaps our human concepts of God are still tied to a previous century. The twenty-first century after Christ may well represent a new Renaissance in human knowledge, a new embarkation into the concepts of the future. Persons now living can hardly imagine the small amount of knowledge and the limited concepts of the cosmos which man had when the scriptures of the five major religions were written. Perhaps old scriptures need new interpretations. The Theology of Humility seeks to build on the great theologies of the past and present and does not oppose any other theology. It welcomes the ideas and inspiring literature of all religions. But perhaps we should be open to the possibility of various new unprecedented religions where the revolutions in our conceptions of time, space and matter significantly shape our theology. Perhaps, while recognizing that God should not be thought of as impersonal, our names for God should be less heavily focused on personhood, since their usage favors man-centered concepts. The Creator seems to be both transcendent and imminently accessible both by science and by prayer, ready to transform the lives of those who invite him in.

The Theology of Humility encourages thinking which is open minded and conclusions which are qualified with the tentative word "maybe." It encourages change and progress and does not resist any advance in the knowledge of God or of nature, but is always ready to rethink what is known and to revise the assumptions and preconceptions behind our knowledge. It is possible that through the gift of free will God allows us to participate in this ongoing creative process. Perhaps a prerequisite on our part is to look beyond our biases and our fears, our personal hopes and aspirations, to see the glorious planning and the infinite majesty of the Planner. Maybe we should also ask ourselves— whether we are students of the natural or the spiritual worlds—to study and experience the ultimate relationships between physical and spiritual realities in our own lives.

These ideas, with their emphasis on creativity and progress, may be illustrated by the following tentative propositions:

1. Maybe it is egotistical to think that progress in religion is not necessary or not possible.
2. Maybe it is self-centered to believe that we stand at the end rather than at the beginning of God's creative process.
3. Maybe it is self-centered to think that we are approaching an omega point rather than ever-increasing diversity beyond imagination.
4. Maybe an evolutionary process without purpose could not create humans dominated by purpose.
5. Maybe God is only just beginning to create his universe and allows each of his children to participate in small ways in this creative evolution.
6. Maybe progress in religion could be more rapid if we could avoid words like "belief" and "think" and "faith" and use instead words like "evidence," "test," "research," "probable," and "empirical."

The third point of the Humility Theology statement in the Bylaws is as follows:

Diversity and Constant Change

The Theology of Humility does not encourage syncretism but rather an understanding of the benefits of diversity. Constant change would seem to be the character of our universe. Despite the cycles of the day and night and the seasons, we may be learning that nothing really repeats. The pattern that unfolds with time may not close back upon itself. Maybe it moves ahead, upward a little at each turn. The evolution of our universe would seem to be vast in its conception, yet curiously experimental and tentative, a truly creative work in progress. Perhaps human beings, so late an appearance in this evolutionary process, have been given some creative role in seeking to understand and interpret awesome and mysterious processes which science only now begins to fathom. We suppose that our part might be likewise to conceptualize and experiment over a wide diversity of possibilities in the physical and spiritual worlds.

In the physical world perhaps we should reexamine the arrow of time. Thermodynamics has tended to provide a picture of irreversible movement from order to randomness as the universe "runs down." But this is the antithesis of what appears to happen in the evolution of the universe and of life. It is as though apparently self-integrated units of the simplest matter exhibit powers which successfully oppose the trend to randomness and produce instead orderly events and structures. Where manmade structures decay, natural systems seem driven toward growth and toward greater diversity and complexity. Here perhaps we have one of the greater laws describing the nature of the cosmos. If only blind chance were involved in evolution, we might expect decay and disorder. Yet the end product of evolution up to this point is a conscious being endowed with a remarkable brain and dominated by the purposes of the Creator.

In the spiritual world perhaps diversity is also reflected in the variety of the religions and in the multiplicity of denominations. It may be that this increasing diversity provides for a freedom and a loving and healthy competition without which there might be only lesser progress. Perhaps we should applaud the new research programs and the new organizations arising in each of the world's religions.

These characteristics of diversity and change may be illustrated by the following tentative propositions:

1. Maybe God is the reality behind all the things seen and the vastly greater abundance of things unseen.
2. Maybe space, time, matter, and energy are a few of the creations or aspects of God that humans are able to comprehend. Maybe multitudes of other creations or aspects are not comprehensible by the human intellect.
3. Maybe the awesome mysteries of magnetism, gravity, light, knowledge, imagination, memory, love, faith, gratitude, and joy are all part of God, and he is much more.

4. Maybe humans who dwell in three dimensions can comprehend only a little part of God's multitude of dimensions.

The fourth part of the Bylaws description of Humility Theology reads:

Encouraging Progress

The Theology of Humility does not rely on man-made institutions or governments, nor does it seek to influence them. Perhaps one of the greatest developments in human history has been the increasing possibility for the freedom of each individual to learn and grow and develop. The Theology of Humility seeks to improve the human condition by internal and spiritual sources rather than by external human governments or institutions. It encourages worship of the Creator rather than dependence upon government, and spiritual growth rather than human, social and political activities.

The Theology of Humility suggests that tremendous benefits could accrue from our greater understanding of spiritual subjects such as love, prayer, meditation, thanksgiving, giving, forgiving and surrender to the Divine will. It further suggests that since science is opening our eyes to the vast works of an infinite Creator, science may also be applied to varieties of experimental and statistical study of these spiritual entities. It may be that we shall see the beginning of a new age of "experimental theology" wherein studies may reveal that there are spiritual laws, universal principles which operate in the spiritual domain just as some natural laws function in the physical realm. Perhaps we will discover that the sphere of the spirit is intensifying as God's evolving plans unfold and accelerate.

Perhaps research foundations and religious institutions should devote vast resources and manpower to these scientific studies in the spiritual realm, equal or greater in magnitude to those currently expended on studies in the physical realm. There could be enormous rewards in terms of increased human peace, harmony, happiness, and productivity if we collected more evidence that the God of the universe has put us here on this planet to learn from and challenge each other and to act as channels to radiate God's love, wisdom, and joy.

These characteristics of Humility Theology may be illustrated by the following tentative propositions:

1. Maybe civilization cannot neglect the humble seeking for God and his purposes.
2. Maybe children cannot be prepared for life by schooling in knowledge that has only a material emphasis.
3. Could vast benefits for humanity result if we began allocating money and human resources for spiritual information and research as we now do for research in the sciences?
4. Is it possible that the intellectual sphere is encompassed by a spiritual sphere, or that possibly spirit could be the underlying and original and continuing source of all creation?
5. Maybe we need a new branch of science about unseen spiritual principles, information, and research.

Humility Theology as a Foundation for Future Research

I am enthusiastic about the possibilities for powerful new insights about meaning and purpose through empirical and statistical studies of spiritual subjects. In an earlier book with Robert Herrmann called *Is God the Only Reality?* we spoke of the new opportunities through Humility Theology:

> When we come to appreciate all these limitations—the extremely thin veneer that is our knowledge of the universe, our limited perspective with respect to the breadth of reality, and our own precarious place in the vast scheme of things, we might well expect a failure of nerve or a crisis of ego. But this is just the point where we ought to be, if we are to experience humility. For the pride and self-esteem that our society so prizes and encourages is the very root of our problem because it ignores our proper relationship to the Creator God who is the Life giver and the Light giver.
>
> In an earlier book we had talked about the Creator of the uni-

verse as the Light giver who seeks to reveal Himself in our universe through intricate design and purpose, and we have talked in this book about God as the Life giver, whose love and faithfulness are revealed in human freedom and natural law—chance and necessity working together to create a world of infinite variety and beauty—form set free to dance. We have been struck with the remarkable contrast between ourselves—minute specks in one tiny star system in one of the smaller galaxies of the known universe and God who brought all this into being. From this vantage point we are impressed with the irony of statements by some small minority of scientists that their ideas of the universe might preclude the need for a Creator. It seems rather like Joseph Bayly's parable of the row computer having a learned discussion about whether man exists! And we are also impressed with how often religions' leaders insist that their interpretation of the truth of God excludes any alternatives. By this means many who could begin the walk of faith are turned away . . .

It is difficult not to see the strong element of pride and self-centeredness in both the atheistic scientist and the religious separatist. Both have excluded the possibility of other valid sources of truth. An attitude of humility would open their minds and hearts, freeing them to greater understanding. It would encourage a spirit of exploration, which for many potentially creative people is no longer considered an option. The arrogance of many scientists and religionists has led to the impression that everything is settled, that there are no frontiers left and no new worlds to conquer. Yet from what we have tried to share about the very limited degree of our present understanding in both science and religion, it should be evident that there is a whole universe to be explored. We have barely scratched the surface! In biological science, for example, the burgeoning fields of neural science and developmental genetics beckon with fascinating questions about the relationship between the mind and the brain, about the developments of the embryo and the mechanisms of macroevolution. In theology, the relationship between mankind and the rest of the created order need fresh study in light of the evidence of the recency of our origin and yet the closeness of our relationship to the rest of living things. Questions raised in behavior genetics and environmental science offer tremendous and exciting challenges for the students of religion.

Scientists and theologians should see opportunity in a theology of humility to pool their resources and explore together the vast reaches of the universe. We take as our models those careful scientists who are not deluded by intellectual pride and who do not deny the metaphysical. Only arrogance would lead others to assert that what they cannot comprehend cannot exist. We see the opportunity for an integration of the disciplines of science and religion. The really paramount question about our nature and the meaning of our pilgrimage are strongly interdisciplinary. Already science is demonstrating the fruitfulness of interdisciplinary studies in the relatively new field of neural science and the very new field of cognitive science. We forecast tremendous advances in human understanding when humble scientists and theologians meet together in joint interdisciplinary research, in a kind of experimental theology.

The future will be open to the scientific exploration of spiritual subjects such as love, prayer, meditation, and thanksgiving. This new exploration may reveal that there are spiritual laws, universal principles that operate in the spiritual domain, just as natural laws operate in the physical realm.

Such an experimental theology would be God-centered, recognizing the inadequacy of our senses and our intellect to comprehend the Creator, yet realizing that God also manifests himself through a spiritual dimension that not only speaks to each of us personally but seems to pervade the world we live in. Such an experimental theology will recognize that a new Renaissance in human knowledge is coming, building on the powerful insights of the past with new data from the physical and human sciences. This future religious emphasis will encourage thinking that is open minded, and conclusions that are tentative. It will encourage diversity rather than syncretism, even as our universe has proved to be in constant change and progressive development.

With this new sense of humility, the way is now open to examine scientifically the spiritual nature of human beings. Indeed, several researchers have already begun an extensive examination of religious variables in health care. Of special significance is the work of psychiatrist David Larson and his colleagues at the National Institute of Healthcare Research of Rockville, Maryland, in revealing the extent to which religious commitment is ignored or mismeasured in clinical medicine and the work of Herbert Benson, author of the best-selling book *The Relaxation Response.* Benson, the director of the Mind-Body Medical Institute at Harvard Medical School and the New England Deaconess Hospital, is conducting research into the relationship between spirituality and health that demonstrates that spiritual experience is correlated with increased life purpose and satisfaction, a health-promoting attitude, and decreased frequency of medical symptoms...

Perhaps the future will see research foundations and religious institutions devoting huge resources to spiritual research just as today we provide enormous amounts of funding for research into physical health. There would be great rewards in terms of increased human peace, harmony, happiness, and productivity as evidence accumulates that mankind's spiritual nature is the central aspect of our being.

Finally, if reality eludes us because it is deeper and more profound than we can ever truly comprehend, because it is an expression of the God of the universe, then it should be our finest goal to know God by every avenue open to us. Surely the best and most profound studies in all of history will come when performed humbly with the expectation of more fully knowing the Creator God and his purposes for his creatures.[3]

Toward a New Field of Science

Joining scientific research with Humility Theology can be the foundation for a new field of science that seeks to learn new aspects of the Creator (spiritual information) by studying the deeper aspects of nature and by proposing new concepts to be verified or falsified. The list of subjects for scientific study could include the following:

- Further evidence for purpose.
- Explanations for the comprehensibility (at least in part) of nature.
- Research into significance of the following scientific observations or inferences:
  1. That the creation is:
     —Mostly unseen.
     —Much larger than previously thought.
     —Apparently much older than previously thought.
     —Increasing and even accelerating in its diversity.
     —Increasing and even accelerating in its complexity.
     —Ruled by law.
  2. That human beings are:
     —Highly creative.
     —Dominated by purpose.
     —Accumulating knowledge at an accelerating rate.
     —Spiritual beings exhibiting love, prayer, thanksgiving, forgiveness, sacrifice, honesty, ethics, etc.
     —Probably not the only self-conscious, intelligent creatures in the universe.

It is my hope that the John Templeton Foundations will make a significant contribution to this spiritual quest through science. The Foundations' involvement has been described most recently in *The Humble Approach,* published in a revised and expanded edition in 1995, as follows:

> To aid in this new search for truth, the John Templeton Foundations, Inc. of America and the Templeton Foundations in other nations have expanded in scope with the formation of a center for the study of Humility Theology with a view to promoting progress in religious thinking. The center has been named the Humility Theology Information Center and has as two of its major components a series of research programs and an advisory board of prominent scientists and theologians interested in

progress in religious thinking. A list of the [current] Advisory Board [can be found at the end of this book].

The research program that the Center has undertaken has three areas of concentration:

1. Utilization of scientific methods in understanding more about the work and purpose of the Creator.
2. Research on studying or stimulating progress in religious information.
3. Research on the benefits of religion.

Examples of projects initiated by the Center include:

- Bibliographic surveys of work by scientists on spiritual subjects.
- Programs to assess the extent of teaching of university and college courses on science and religion together and to stimulate courses emphasizing progress in religion.
- Training modules on religion and on psychiatry that illustrates the extent to which spiritual factors may influence clinical therapy.
- Programs to encourage scientists and theologians to publish papers on humility theology.
- Programs of lectures on the relationship between science and theology presented at universities and colleges in North America and Europe and, more recently, at large churches in the United States.[4]

It is the Foundations' hope that these questions concerning God's ultimate relation to the visible world may promote greater interest on the part of scholars and researchers in the exploration of theological and philosophical implications of the momentous and accelerating scientific discoveries of our time. Nothing could be more important than to study our true relation to the awesome and wonderful Creator of the universe. For my own part, I continue to ask God, as I have in my daily prayers for the past ten years, to "open my mind and heart more fully to receive thy unlimited love and wisdom, and to radiate these to thy other children …"

How Large Is God?

In the chapters that follow, nine distinguished scholars and scientists who have served on the Board of Advisors of the John Templeton Foundation present a variety of views on the dimensions of God. In Part 1, Perspectives on Science and Theology, an astronomical perspective is presented in Harvard professor Owen Gingerich's story of how our view of the universe, and hence its Creator, has grown to immense proportions in the scientific age. In the second chapter, physicist Freeman Dyson, of Princeton's Institute for Advanced Study, explores for us the border between science and religion, and concludes that "God is in the richness of the phenomena; the science is ours."

The next four chapters provide, in Part 2, views of how we approach an answer to the question, How large is God? Another physicist, Professor Russell Stannard of Britain's Open University, shows us yet another view of science and religion as sources of truth; both scientists and theologians have learned to live with paradox, seeming contradiction that leads to deeper truths. Next is a contribution of Harvard Medical School's Herbert Benson and Marg Stark. Dr. Benson's long study of the relationships between mind and body has led him to the conclusion that faith and its remarkable therapeutic value may be "wired in" by our Creator.

The last two chapters in this section take sweeping views of the current scientific data about the universe. Professor Howard Van Till, physicist-cosmologist at Calvin College, emphasizes the stupendous creativity resident in the matter and material systems of a universe that must have an awesome Creator as its "giver of being." Physicist and theologian Robert J. Russell, director of the Center for Theology and the Natural Sciences and professor of theology and science in residence at the Graduate Theological Union in Berkeley, provides deep insights from big bang cosmology and from the nature of mathematical infinity about "the God who infinitely transcends infinity."

Part 3, Limits to Scientific and Theological Answers, begins with a chapter by theologian Martin Marty, Fairfax M. Cone Distinguished Service Professor at the University of Chicago and founder and senior scholar at the Park Ridge Center for the Study of Health, Faith and

Ethics. His burden is to explore the predicament theologians and humanists face in trying to make meaningful scientific statements when the language of science is mathematics and the essential feature is quantification. Next, physicist-astronomer John Barrow, of the Astronomy Centre at the University of Sussex, adds a chapter on the limitations scientists face in arriving at anything remotely resembling ultimate answers. He concludes, "There are intertwined logical structures that underwrite the nature of reality." Finally, the last chapter, by biochemist Robert Herrmann, on the faculty in chemistry at Gordon College, provides several examples from current science on the elusive nature of ultimate reality. In the end, he concludes that asking questions about the universe and its Creator leads to even more profound questions, for "reality is deeper than we are."

Notes

1. Bylaws (Sewanee, Tenn.: The John Templeton Foundation, 1987).
2. "Bridging Two Worlds—An Interview with Sir John Templeton, *Second Opinion*, July 1993.
3. John M. Templeton and Robert L. Herrmann, *Is God the Only Reality?* (New York: Continuum Publishing Group, 1994) pp.166-170.
4. John M. Templeton, *The Humble Approach: Scientists Discover God,* 2nd ed. (New York: Continuum Publishing Group, 1995) pp.130-131.

PART 1

*Perspectives on Science
and Theology*

An Astronomical Perspective

OWEN GINGERICH

The Human Scale

> *Once upon a time all the world spoke a single language and used the same words. As men journeyed to the east, they came upon a plain in the land of Shinar and settled there. They said to one another, "Come, let us mold bricks and burn them hard." Then they said, "Come, let us build ourselves a city and a tower with its top in the heavens, and make a name for ourselves."*—GENESIS 11:1–4[1]

Whatever the implications of the Genesis story for the clash of rural versus urban life, the image of the world's first skyscraper remains a vivid vignette. What inspired the text was not monumental architecture in Babylon, which came several centuries after this part of Genesis was first written, but a prior Akkadian literary tradition that reads, "The first year they molded its bricks, and when the second year arrived they raised the head of Esagila toward Apsu," that is, toward the unbounded heavens.[2] The notion that the men of Shinar or Akkad could build a tower with bricks and mortar reaching into the heavens now seems almost quaintly naive. Today, we know that even the tallest mountains on earth hardly create a wrinkle on the smooth spherical shape of our planet when viewed, let us say, from nearby Mars, and Mars is only the tiniest first step in our expanded vision of what "the unbounded sky" means for us.

In contrast, all the ancient literatures, including the Bible, portray a stage of intimate proportions. God is near at hand, whether in Eden or on Sinai. When Elisha or Christ ascended into heaven, the trip to heaven was not envisioned as a lengthy cosmic journey. Even the pagan gods of the Greeks, ensconced on Mount Olympus, were not all that remote.

There is no great cosmic scale in antiquity. "Thou hast stretched out the heavens like a tent," exclaims the Psalmist in a human-sized metaphor. In the heavens, a tent is fixed for the sun, declares Psalm 19. Early in the fourth century, Lactantius Firmianus, the church father known as the Christian Cicero, went so far as to use these scriptures to argue that the earth was flat, although he definitely remained in the minority.

In a series of brilliantly encyclopedic treatises, Aristotle, the fourth-century B.C. Athenian philosopher, laid out a vision of a spherical earth in the center of a cosmos surrounded by weightless, eternally rotating shells that carried the planets and stars, with an outer "primum mobile," or prime mover, kept in motion by God's goodness.[3] As for the details of the motions or the dimensions of the spheres, Aristotle had hardly any information. The contrast between his *De coelo* or his *Metaphysics* and the thoroughly numerical treatise by the Alexandrian astronomer Claudius Ptolemy (c. A.D. 150) is truly striking.

In the five centuries between Aristotle and Ptolemy, other Greek philosophers had conceived ingenious methods to deduce not only the size of the Earth but even the distance to the Moon, approximately 60 earth-radii away. Beyond the Moon lay the other wandering stars: Mercury, Venus, the Sun, Mars, Jupiter, and Saturn. A clever but highly faulty scheme for getting a distance to the Sun yielded a result of nineteen times farther than the Moon, or (19 x 60 =) 1,140 earth-radii. This number held sway until the end of the sixteenth century, well past the time of Copernicus. Working with this spurious dimension (approximately twenty times too small), Ptolemy and his successors carefully nested the mechanisms for the planets one next to another, finally placing the spherical shell of fixed stars just beyond Saturn, to give a total distance of 20,000 earth-radii from earth to sky.

While a universe 80 million miles in diameter must have seemed staggeringly large to the men of the Renaissance, by late twentieth-century standards the medieval universe they inherited seemed downright cozy in scale. And indeed, fifteenth-century artists, authors, and book illustrators made commonplace a vision of the geocentric cosmos with the earth surrounded by the crystalline planetary spheres and all surmounted by God's empyrean domain, populated by angels, apostles, musicians, and the elect.

Among the notable images of this genre was a woodblock by Michael Wolgemuth, assisted by the young apprentice Albrecht Dürer, that appears as a full-page display in the well-illustrated *Nuremberg Chronicle,* the great coffee-table book of 1493 (see figure 1). A similar image, Pierro di Puccio's wall-sized fresco, is found in the Camposanto next to the cathedral and Leaning Tower in Pisa. And yet another, Giovanni di Paolo's small oil painting, is today in the Lehman collection at the Metropolitan Museum of Art. These Christianized versions of the Aristotelian cosmos had become well established as the Western view of sacred geography in the early 1500s.

As for Columbus, there was no need for him to argue at the Spanish court that the world was round. That much was already granted, and the recovery of Aristotelian texts in the fourteenth century had reinforced the arguments for a spherical earth. Instead, Columbus' famous quarrel with the clerics and scholars of the University of Salamanca had to do with the size of the earth. Ironically, his error in badly underestimating the earth's circumference made his voyage seem feasible, and to the end of his life Columbus was unwilling to admit that he had not sailed all the way to the Orient. The Spanish scholars were pretty nearly right about the size of the earth, and they, knowing the formidable distance west to Japan, definitely would not have made the trip.

While Columbus did not directly reset the dimensions of the world, the startling newness of his discoveries, and the realization that the ancients' knowledge was enormously incomplete, helped open Western Christendom to fresh ways of thinking that eventually led to a complete rescaling of the universe.

The Astronomical Scale

The century following the fall of Constantinople (in 1453) brought immense changes to Western Civilization and Western Christendom. These included the rise of printing, the voyages of exploration and the discovery of the New World, and the Protestant Reformation. And they also included, like a time-bomb ticking away, the publication in 1543 of Nicolaus Copernicus' *De revolutionibus orbium coelestium.*

De lanctificatione leptime diei

Onfummato igitur mundo: per fabzicam diuine folercie fex dierum. Creati em difpofiti z oznati tandē pfecti funt celi z terra. Compleuit ꝙ glioſus opus ſuū: z requienit die feptimo ab operib⁹ manuum ſuarū: poſtꝗ cūctum mundū: z omnia que in eo funt creaſſet: nō quaſi operando laſſus: ſed nouam creaturam facere ceſſauit: cuius materia vel ſimilitudo non pceſſerit. Opus enim pzopaga⸗ tionis operari non definit. Et dominus eidem diei benedixit: z ſanctificauit illū: vocauitꝗ ipfum Sabati quod nomem hebzaica lingua requiem ſignificat. Eo ꝙ in ipfo ceſſauerat ab omi opere qd patrarat. Ʒn z Ʒudei eo die a labozibus pzopzijs vacare dignofcūtur. Quem z ante leges certe gentes celebzem obfer uarunt. Ʒamꝗ ad calcem ventum eſt operum diuinozum. Ʒllum ergo timeamus: amemus: z veneremur. Ʒn quo funt omnia ſiue viſibilia ſiue inuiſibilia. Et a domino celi: domino bonoʒ omniū. Cui data omis poteſtas in celo z in terra. Et pzeſentia bona: quatenus bona ſint. Et veram eterne vite felicitatem quera⸗ mus.

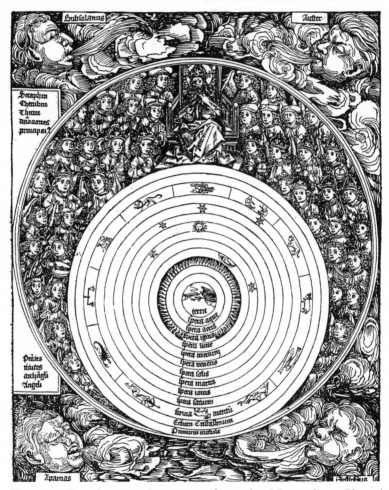

FIGURE 1—Geocentric cosmos from the *Nuremberg Chronicle* (1493).

Published only days before his death, Copernicus' book made relatively little splash in intellectual circles despite its radical reworking of the sacred geography. The Polish astronomer took Earth out of its time-honored central position and proposed that it traveled swiftly around the Sun in an annual orbit. But the book was almost universally viewed merely as a geometrical scheme, "not necessarily true nor even probable," as an anonymous introduction put it. "You may have heard about the hypotheses in this book and are concerned that all of liberal arts are about to be thrown into confusion. But don't worry about it. A philosopher will seek after truth, but an astronomer will just take what is simplest. So don't expect to find truth here, lest you leave a greater fool than when you entered." This warning and disclaimer had been silently added by Andreas Osiander, the Lutheran clergyman who served as proofreader for the book, and for a while it protected Copernicus' work from strictures by Protestants and Catholics alike. Not until seven decades later, after Kepler and Galileo had begun to treat the heliocentric arrangement as a real description of the cosmos, did *De revolutionibus* begin to have a revolutionary impact, and then it was finally proscribed and listed on the *Index of Prohibited Books*.

Copernicus himself was a staunch Catholic and a canon at the Frombork (Frauenberg) Cathedral in the northernmost diocese of Poland. Undoubtedly, he worried about potential objections from churchmen who had settled comfortably into an Aristotelian view of the earth and heavens, and this may have contributed to his reluctance to publish his manuscript. But to forestall criticism from ecclesiastical quarters, he included in his dedicatory letter to the Pope that "mathematics is for mathematicians," implying that technical astronomy is for astronomers and not clerics or theologians.

When Copernicus proposed his heliocentric theory, he knew that objections could be raised because of the lack of a tiny observed annual parallactic motion of the stars, which would be caused by the motion of the Earth. Copernicus pointed out that his system actually explained, for the first time, why the so-called annual retrograde motion of Jupiter was smaller than that of Mars (which orbited closer to Earth), and why the retrograde motion of Saturn was less than that of Jupiter. The stars' annual motions were still less imperceptible, because they were so far away. "So vast, without any question, is the Divine handiwork of the almighty Creator," he declared at the end of his stirring cosmological chapter. Curiously enough, this sentence was specifically marked for censorship when *De revolutionibus* was placed on the *Index*. Why excise such a pious statement about the vastness of God's creative powers? Apparently, the Inquisitors thought Copernicus' rhetorical flourish had too much of a whiff of reality for a scheme they believed was only hypothetical.

Copernicus, like Aristarchus before him, had conceived of a starry sphere large enough to conceal any annual parallactic motion. Interestingly, Copernicus' actual planetary system was *smaller* than Ptolemy's, because of the economic centering of all the planetary spheres on the Sun, but the starry frame was much larger, perhaps 200 times as great in diameter. Copernicus placed all the stars at the same distance on the surface of the starry sphere; like most medievals he assumed that paradise lay immediately beyond. In principle, Copernicus had not shattered the tidy medieval cosmos; he had simply made it much larger.

The English astronomer Thomas Digges was the first to depict the stars not attached to a single spherical shell but spread out through space. He appended a translation of the short cosmological part of Copernicus' *De revolutionibus* to the 1576 edition of his father's perpetual almanac and included a fold-out diagram of the heliocentric system (figure 2). How did Digges cope with the divine geography of heaven? He labeled his stars, "This orbe of starres fixed [i.e., fixed stars] infinitely up extendeth hit self in altitude sphericallye, and therefore immovable, the pallace of foelicitye garnished with perpetual shininge glorious lightes innumerable, farr excellinge our sonne both in quantitye and qualitye, the very court of coelestial angelles, devoid of greefe and replenished with perfite endlesse joye, the habitacle for the elect."[4]

Thus, it was Digges who took the first dramatic step toward an infinite universe, but he nicely accommodated his cosmos to include a place for paradise among the stars themselves, just as his father's earlier diagram of the ancient geocentric system included a label around its outer bound reading "Here the Learned do place the Habitacle for the Elect."

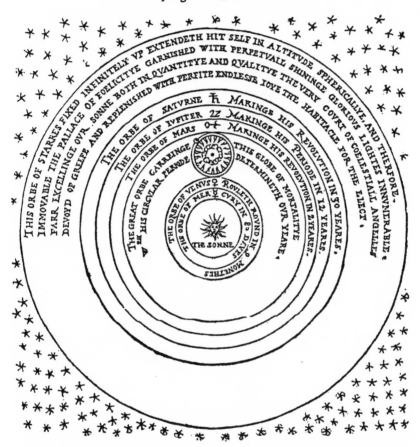

FIGURE 2—Thomas Digges' diagram of the heliocentric system shows, for the first time, that the stars are distributed throughout space.

Not until six decades after *De revolutionibus* was placed on the *Index* did astronomers begin to get a real clue about how distant the stars really are. Most leading scientists by then accepted the Copernican system as the real arrangement of the planets, despite the lack of any observed annual stellar parallax. In fact, the stars are so distant that the annual angular displacement of the closest ones are similar to the angular size of a dime measured two miles away. In the absence of any instruments precise enough to measure such a small quantity, astronomers turned to another method for finding the distances to stars, which depends on the known geometric diminution of light with distance, the so-called "faintness means farness" method.

Perhaps the most famous early use of the faintness means farness method was given in 1698 by Christiaan Huygens in his *Cosmotheros* or in English, *The Celestial Worlds Discover'd*. Huygens describes how he reduced incident sunlight geometrically until its brilliance matched his memory of the luminance of the bright star Sirius. In this way he got a ratio of 27,664 for the distance of Sirius compared with that of the Sun. "And what an incredible distance that is," he exclaims.

> For if 25 years are required for a bullet out of a Cannon, with its utmost Swiftness, to travel from the Sun to us; then by multiplying the number 27664 into 25, we shall find that such a Bullet would spend almost seven hundred thousand years in its Journey between us and the nearest of the fix'd Stars. And yet when in a clear night we look upon them, we cannot think them above some few miles over our heads. What I have here enquir'd into, is concerning the nearest of them. And what a prodigious number must there be besides of those which are placed so deep in the vast spaces of Heaven, as to be as remote from these as these are from the Sun! For if with our bare Eye we can observe above a thousand, and with a Telescope can discover ten or twenty times as many; what bounds of number must we set to those which are out of reach even of these Assistances! especially if we consider the infinite Power of God. Really, when I have bin reflecting thus with my self, methoughts all our Arithmetick was nothing, and we are vers'd but in the very Rudiments of Numbers, in comparison of this great Sum.[5]

Although Huygens (like Ptolemy in his ancient estimate of the Sun's distance) was off by a factor of 20, by the second half of the eighteenth century, he and his contemporaries had neatly bracketed the distance of Sirius by analyzing the apparent brightness of that star and by making the risky assumption that it was intrinsically similar to the sun. Had Huygens tried instead the nearby star Rigel, the result would have been wide of the mark, although it would have given an approximately correct result for the nearest stars. Indeed, as he explained, his answer referred to a relatively nearby star; how much farther did the stellar system actually extend?

Nearly a century would pass before this question would be addressed. William Herschel, who had come to England from Germany as an itinerant musician, became a passionate amateur astronomer, but his musical profession and his amateur astronomical standing were almost literally destroyed overnight on March 13, 1782, when he discovered the planet Uranus. With royal patronage, he was able to devote himself entirely to astronomy and to build ever-larger telescopes. With them he undertook an ambitious gauging of the heavens, counting stars in various directions to determine the relative extent of the immense Milky Way star cloud in which the Sun was situated.

Herschel knew that the great milky band of light girdling the sky was actually formed by the confluence of numerous stars too faint to be seen individually but readily visible in his huge reflectors. He suspected that these stars were arranged in a giant disk, with the Sun as one star near the center. When the line of sight ran edgewise through the thin disk, the multitude of stars crowded together to give the impression of the Milky Way. Herschel further assumed that within the Milky Way the stars were distributed uniformly right up to the edge of the disk, so that if he simply counted all the stars in some small part of the sky, he could then reckon the relative extent of the Milky Way in the direction.

While he could easily confirm the shape of the Milky Way, he had no clear way to establish the actual distances involved. "And though my single endeavours should not succeed in a work that seems to require the joint effort of every astronomer," he wrote, "yet so much we may venture to hope, that, by applying ourselves with all our powers to the improvement of telescopes, which I look upon as yet in their infant state, and turning them with assiduity to the study of the heavens, we shall in time obtain some faint knowledge of, and perhaps be able to partly delineate, *the Interior Construction of the Universe.*"[6]

A child of the Enlightenment, Herschel, unlike Isaac Newton, Christiaan Huygens, and the astronomers before them, did not pause to reflect on the power and creativity of the Almighty. Earlier, in speaking before the Bath Philosophical Society, he had noted that "Space is the stage on which Omnipotence displays its wonders."[7] But in his papers for the Royal Society, it would seem that a watershed had been passed, and astronomers no longer felt obliged to display a paean of praise to God in describing marvelous aspects of the cosmos.

In attempting to pin a distance scale to his picture of the Milky Way, Herschel did not use the faintness means farness principle directly. One reason was the great difficulty in measuring the stellar brightnesses quantitatively, especially for the fainter and presumably more distant stars. After yet another century passed, photography had come to the rescue, for the brightnesses of stars could be reliably measured on the photographs. In the hands of the Dutch astronomer Jacobus Cornelius Kapteyn, the growing abundance of photographic data finally yielded a numerical answer to the extent of the Milky Way disk. His numbers showed the Sun in the center of a system roughly 300,000,000,000,000,000 kilometers in diameter—an immense size, far beyond anything seriously considered in astronomy before 1900— and a number so large that clearly other units of measurement were required. Astronomers settled on light-years, the distance light travels in a year. This made the "Kapteyn universe" 30,000 light-years in diameter.

By the time Kapteyn published his final results in 1923, astronomers had already begun to consider still larger distances. In 1918, several researchers had photographed very faint new stars, or *novae*, in the Andromeda Nebula, a fuzzy but quite faint spiral nebula stretching across three degrees of the sky in the constellation Andromeda (that is, equivalent to stretching over half the distance between the pointer stars of the Big Dipper). If these novae matched those in our Milky Way, this put the Andromeda Nebula a million light-years away. It would be truly a spiral galaxy, an island universe far beyond the bounds of our Milky Way's disk.

At the cutting edge of science the results are often contradictory and murky, and so it was with the novae in the Andromeda Nebula. A much, much brighter nova had been spotted near the nucleus of the nebula in 1885, and if that had been a typical nova like those nearby in the Milky Way, then the Andromeda Nebula would lie much closer, in the immediate neighborhood of the Milky Way itself. The confusion was compounded by new results from Harlow Shapley, a brash young American astronomer working with the world's largest active telescope, a 60-inch reflector on Mount Wilson, just north of Pasadena, California.

Quite unexpectedly, Shapley's seemingly unrelated field of research—variable stars in globular clusters—led to a new picture of the Milky Way far different from Herschel's or Kapteyn's. And once again, the principle of faintness means farness was intimately involved. The globular clusters, Shapley's chosen topic, are immense congeries of tens of thousands of stars in relatively compact spherical swarms. From the tropics, a few globular clusters can be seen with the naked eye, and from north temperate latitudes several can be found with a good pair of binoculars. Altogether the Milky Way includes about a hundred globular clusters. Shapley noticed that about thirty of these clusters were concentrated in only five percent of the sky, near the constellation Sagittarius in the summer southern Milky Way. His challenge was to find their distances.

Among the most luminous stars in these globulars are a few that pulsate in brightness with periods ranging from a day or so to a couple of months; they were recognizable as a special type, named cepheid variables after their prototype in the constellation Cepheus. From a dozen relatively nearby cepheids, Shapley was able to deduce their high intrinsic brightnesses, and then from the apparent faintness of the examples in several representative globular clusters, he could calculate the farness of the globulars. At the time, the results of his measurements seemed quite staggering. The typical cluster seemed to be about 50,000 light-years away. Shapley then made the daring conjecture that the globulars were centered about a distant and invisible nucleus of the Milky Way, which lay about 50,000 light-years in the direction of Sagittarius. By symmetry considerations, he inferred that the entire Milky Way system was well over 200,000 light-years in diameter.

The older astronomers, steeped in the Kapteyn tradition of star counting and stellar statistics, were not about to accept Shapley's radical new hypothesis without a fight. They challenged his calibration of the cepheid variables and his bold extrapolation to an unseen nucleus. But gradually the evidence accumulated, and eventually the debate on the scale of the Milky Way seemed settled. There remained, however, the glaring discrepancy between Shapley's picture of the Milky Way and the comparatively smaller, sun-centered Kapteyn universe. Until 1930 astronomers maintained an uneasy truce; in that year Robert Trumpler, an astronomer working at Lick Observatory in California, made the unexpected discovery that space within the Milky Way is not fully transparent. While astronomers had known of the existence of large cosmic dust clouds, they had not realized that seemingly clear lines of sight also were plagued by interstellar dust, and, in effect, Kapteyn had been surveying a dusty universe. The edges of his system were not the boundaries of high stellar densities but were the artificial edges caused by a dusty fog.

As for the spiral nebulae, their status was settled somewhat more quickly. Shapley noticed that the Andromeda Nebula, even at a distance of a million light-years, would not be comparable to the Milky Way. For this and a variety of other reasons (including the nova of 1885), Shapley argued that the spiral nebulae must be relatively nearby and somehow related to our own stellar system. Within just a few years, his rival at Mount Wilson Observatory, Edwin Hubble, used the new 100-inch telescope and Shapley's own calibration of the cepheid variable stars to prove him dead wrong. By taking scores of photographs of the Andromeda Nebula, Hubble discovered cepheid variables there, and once more the faintness means farness rule provided the key. The results, announced at a meeting on New Year's Day in 1925, confirmed the earlier estimates based on the novae: a distance of about a million light-years. As for the anomalously bright nova in 1885, it turned out to be a totally different phenomenon, now called a *super-*

nova. The Andromeda Nebula became the Andromeda galaxy, an island universe quite separate from our own Milky Way galaxy.

Unfortunately, the high luminosity cepheid variables were near the limit of visibility at the distance for the Andromeda Nebula, and in more distant galaxies they were essentially too faint to be seen. To extend the distance scale further, Hubble still had to rely either on the faintness means farness criterion or on the very similar tool, "smallness means distance," not using single stars but entire galaxies and by assuming that these galaxies had similar total luminosities or sizes. The actual process was somewhat more subtle, of course. For example, single, isolated galaxies have an enormous range of intrinsic brightnesses, so Hubble often chose to use clusters of galaxies where a "typical" galaxy could be discerned with more reliability.

Employing these procedures, Hubble was able to estimate the distances to a group of galaxies bright enough for their spectra to be examined. Because a prism spreads out the light into a short rainbow, each colored section is necessarily fainter than the combined white light of the galaxy and, therefore, that much harder to photograph. However, the large telescopes in the American Southwest and especially on Mount Wilson made such a research program possible. The reason that the spectra were particularly interesting was that they showed that the galaxies were moving at high speeds (indicated by a Doppler shift in the color of the spectrum), typically ten or even a hundred times faster than typical stars in the Milky Way. (If Milky Way stars had such high velocities, they would quickly escape from our galaxy.)

Hubble had a big surprise waiting for him. When he arranged the galaxies in order of distance, he found that the farther they were, the faster they were rushing away from us. At twice the distance, the speed of recession was twice as great. This is the signature of a monumental explosion. Consider two projectiles, thrown out at the same moment from some enormous cataclysm: after some time, the one going twice as fast will be twice as far away. Hubble's observational discovery, announced in 1929, described the expanding universe leading from an initial burst now commonly called the big bang.

"There is no way to express that explosion," writes the poet Robinson Jeffers.

> "All that exists
> Roars into flame, the tortured fragments rush away from each
>   other into all the sky, new universes
> Jewel the black breast of night; and far off the outer nebulae like
>   charging spearmen
> Invade emptiness."[8]

Given the distance and rate of recession of a galaxy, it is easy to calculate the time when the big bang took place. To the geologists, Hubble's age of a billion years seemed much too small. By 1929, there were good reasons to believe that the age of the earth was at least several times greater than a billion years. While astronomers had been busily enlarging the distance scale of the heavens, geologists had been at work revising the biblical picture of the age of earth and the antiquity of humankind. Because the enlargement of the time dimension has had just as much impact on our conception of the size of God as has the enlargement of the spatial dimensions, let us now turn to what has been called "the discovery of time."

The Geological Scale

Unlike Aristotle's view, that time is endless and the earth has been here forever, the biblical picture was of a world and time created so recently that its age could be reckoned in human generations. Like the spatial scale of the celestial spheres, the entire biblical span of time took on comfortably human dimensions. Nevertheless, folk in the European Middle Ages had not much sense of the progression of time. Everything seemed static and bound into repeated cycles. However, the notion that things were once quite different was demonstrated rather strikingly and unexpectedly in the fifteenth century. Two scholars, Nicholas of Cusa and Lorenzo Valla, showed that the *Donation of Constantine*, a famous document purporting to give the Roman church primacy over Constantinople, took for granted political structures of the Middle Ages that simply did not exist at the time of Emperor Constantine. These anachronisms proved that the document was a forgery and gave dramatic evidence that things were not as unchanging as they naively appeared.

Three centuries later, the French biologist and natural philosopher Count de Buffon, author of the encyclopedic *Natural History*, boldly proposed that Earth itself would have a very long history, at the very least 168,000 years (and privately his estimate was nearer half a million years). The days of creation as recorded in Genesis, he maintained, were actually epochs of indeterminate length, ranging from 3,000 to 35,000 years. His timetable was wrong, but he had broken the barrier in thinking about the age of the Earth. In the decades that followed, geologists gradually built up the concept of the geological column, the deep piles of ordered sedimentary strata that gave evidence for long and slow deposition of rock layers. Like the astronomer William Herschel, who could delineate the shape of the Milky Way without placing an absolute scale on it, nineteenth-century geologists could arrange the rock layers but could only guesstimate their real ages.

By the time Charles Darwin published *On the Origin of Species* in 1859, not only had the geological column been well established but also evidence was emerging that the different groups of animals characterized successive epochs on the Earth. The more complex creatures were found only in the more recent fossil layers. But how much time did Darwin have at his disposal for the gradual evolution of the species? This was a highly debatable matter. Darwin recognized that for his theory to be credible, he had to convince his readers that the Earth's surface was very old, and so he devoted an important part of his treatise to this point. In 1859, no one had a very clear idea of precisely how old the various rock layers are—that was a major achievement of the twentieth century—so Darwin had to guess that the process he outlined would take place within several billion years.

What Darwin subsequently recognized as the "gravest objection" to his theory was raised not by a biologist or by a churchman, but by the formidable physicist Sir William Thomson (later to become Lord Kelvin). Kelvin set out several theoretical arguments. One depended on the cooling time of the Earth. Another was the time required for the Sun to collapse to its present size (and Kelvin supposed that the Sun got its energy from a slow, continuing collapse). All his estimates led to an age of the Earth of probably not in excess of 100 million years, and with a habitable environment of only a few million years. Darwin retreated but only slightly; in Darwin's lifetime, Kelvin's arguments constituted a serious obstacle to his evolutionary picture. The tension between the geologists and physicists remained unresolved, and in 1892 Kelvin repeated the exact words he had used almost forty years before:

> Within a finite period of time past the Earth must have been, and within a finite period of time to come must again be, unfit for the habitation of man as at present constituted, unless operations have been and are to be performed which are impossible under the laws going on at present in the material world.[9]

The solution to the puzzle Kelvin proposed came not from the geologists but from a physicist, Ernest Rutherford, whose researches on radioactivity removed Kelvin's principal objections. Rutherford's own lively account of the resolution, which came at a Friday Evening Discourse at the Royal Institution in London in 1904, runs as follows:

> I came into the room, which was half dark, and presently spotted Lord Kelvin in the audience and realized that I was in for trouble at the last part of my speech dealing with the age of the Earth where my views conflicted with his. To my relief Kelvin fell fast asleep but, as I came to the important point, I saw the old bird sit up, open an eye and cock a baleful glance at me! Then a sudden inspiration came and I said that Lord Kelvin had limited the age of the Earth *provided no new source was discovered.* "That prophetic utterance refers to what we are considering tonight, radium!" Behold! the old boy beamed upon me![10]

In effect, the energy within atoms, as evidenced by their radioactive decay, vitiated Kelvin's calculations for the cooling of the Earth, and provided a source of energy for the Sun's long-term radiation. And within a few decades, the radioactive decay of uranium and thorium would provide the clocks for measuring the age of igneous rocks, the lavas of ancient volcanoes that were here and there interspersed between strata in the geological column. This placed the onset of the Cambrian period, where the first complex life appears in the fossil record, at about 550 million years ago. Eventually, still older igneous rocks were found, in Australia and in Greenland, and these push the date for the Earth's continental areas back to 3.26 billion years ago.

The oldest measured part of the solar system, however, is not found on the Earth, which in its earliest period was heavily reworked by impacting planetesimals, including the particularly large collider that gave rise to the Moon. Instead, the oldest measured rocks are from a class of carbonaceous chondrite meteorites, of which the Allende Meteorite is perhaps the best example. A ton of this material fell near a small Mexican village of Puebla de Allende in 1969, and fragments have been dated to 4.6 billion years.

These huge time spans far exceed the expansion time of a billion years that Hubble calculated in the 1930s for the age of the universe. Cosmologists jumped into the fray to propose more complicated pictures of the expansion, with long periods of stasis when the universe aged but did not become bigger. A much better solution to the discrepancy came in 1952 when Mount Wilson and Palomar astronomer Walter Baade found the Achilles heel in Shapley's earlier calibration of the cepheid variable stars. It turned out that a second, brighter family of pulsating variables masqueraded as the fainter, random cepheids in our part of the Milky Way, which Shapley had used as his standards. Furthermore, the photographic photometry of the faint stars used for measuring the brightness of the cepheids in the Andromeda galaxy had made their apparent magnitudes systematically too bright. With the advent of powerful photoelectric detectors, the errors were corrected, and the distance of the Andromeda galaxy was more than doubled, while the fainter ones beyond it were corrected even more. This nudged the expansion age of the universe up into the 10-20 billion year range, safely above the ages of terrestrial rocks or solar system meteorites.

Nevertheless, in the decades following 1952, an uncertainty of approximately a factor of two remained in the expansion age. During this time, electronic computers arrived on the scene, making it possible for astronomers to trace the evolutionary history of stars by calculating models of the stellar interiors and the nuclear processes that fueled the stars. While it was difficult if not impossible to calculate convincingly the age of a single isolated star, the theorists had notable success in working out the ages of huge assemblages such as the globular clusters, in which there are samples of many different initial masses all collected together. These ages came out typically around 15 billion years. Depending on which end of the uncertainty the final value of the Hubble expansion falls, the two different methods for finding the age of the universe (from the evolution of star clusters or from the expansion) would either dovetail nicely, or the clusters would turn out to be anomalously (and impossibly) older than the universe.

How can the expansion results be established more securely? Ever since Hubble's discovery of the expansion of the universe, astronomers wished that they could see cepheids at even greater distances than just the Andromeda Nebula and its closer neighbors in space. For example, the initial steps of the distance ladder would be considerably strengthened if cepheids could be detected in the great cluster of galaxies that lies perhaps fifty times farther than the Andromeda Nebula in the direction of the constellation Virgo. But the farness of the cluster guaranteed the faintness of the cepheids, and even if their light could be detected with long exposures from earthbound telescopes, the distant images would generally be confused with adjacent stars because of the prismatic roiling of the earth's own atmosphere. One of the major goals of the Hubble Space Telescope was to provide the crisp sharpness, above the earth's turbulent air, required to ferret out the cepheids in galaxies of the Virgo cluster. This objective was finally achieved in 1995, when researchers found a dozen examples in the Virgo galaxy M100. If M100 is representative of the core of the Virgo cluster (and there are some reasons for thinking that it may not be), then the expansion age works out aggravatingly smaller than the ages of the globular clusters.

Some cosmologists have gone back to the drawing boards, once again to consider that the history of expansion could provide a long interval of stasis, allowing the globular clusters to age while the universe stood still (comparatively speaking!). Other astronomers believe that the discrepancy, less than a factor of two between entirely different physical methods, simply indicates that the whole process is robust and that fine-tuning will eventually resolve the conflicts. While the popular press might try to convince the public that the big bang model of the universe is on the brink of collapse, none of the solutions taken seriously by the great majority of the astronomical community include scrapping it.

One thing seems entirely certain: We live in a vast and very old universe, but a universe that has a history, including a sudden, cataclysmic birth and a slow, majestic evolution as the elements necessary for life on Earth were gradually formed in the giant cauldrons of stellar interiors. A small minority of astronomers envision a universe that has gone on forever, but even they must concede that our part of the universe—the observable part—gives every evidence of a creation at a particular moment in time past. Without this scenario, scientists are unable to understand the observed distribution of atomic elements in the cosmos.

Beyond Scale

When Nikita Khrushchev gleefully reported that his cosmonauts had failed to find heaven in their global circumnavigations, he was lampooning a sacred geography that had long since been abandoned by theologians and laity alike. But where is God in this vast universe as we now view it? Certainly we can no longer envision God at the fringes of the universe actually spying on us, an Old One who neither slumbers nor sleeps.

Furthermore, popular writers frequently remind us that if all geological time (roughly five billion years) is mapped into the calendar of a single year, then the dinosaurs, having been around since mid-December, go extinct just after Christmas; hominoids make their appearance about eight hours before midnight on December 31; and modern *Homo sapiens* comes on the scene a few seconds before midnight. Of what consequence are we tiny mites who occupy such a brief instant in the sea of time and space?

The notion that there might be far more to the universe than meets the eye, and that God could be quite close at hand despite the immensity of the universe, got a powerful boost in 1884 when Edwin A. Abbott published his delightful, exasperatingly sexist, and yet profound fantasy, *Flatland, a Romance of Many Dimensions.* His imaginary two-dimensional world, Flatland, is visited by a sphere from three-dimensional space that, by moving up and down in three dimensions, goes from a point to a large circle and back again as far as the intimidated creatures of two-dimensional Flatland are concerned. At the story's climax, the author, a square, eagerly asks the sphere to show him a four-dimensional or five-dimensional world, for which audacity he is promptly punished and eventually jailed. By then, Abbott's audience is thoroughly convinced that there might just be such a world, even though we cannot easily envision it.

There is something almost romantic about the fact that the human mind (the most complex single entity that we know about apart from God) can not only envision something as much outside itself and its own time frame as a distant big bang but can also project itself into unobserved dimensions as intimately close as anything we can imagine. In very recent times, cosmologists have begun serious consideration of an origin of the universe with ten space-time dimensions, six of which never expanded and are effectively rolled up and tucked in so that we cannot sense them. Who knows what goes on in these other worlds that we can only dimly imagine!

The theme of an unseen world around us figured centrally in the philosophy of Sir Arthur Eddington, perhaps the century's most gifted astrophysical theorist. A Quaker by birth and inclination, he gave a famous lecture at the 1929 Friends Yearly Meeting in London on the topic of Science and the Unseen World. For Eddington, the unseen world was rather different from higher dimensions invisible to our constrained senses. Rather, it was the mostly empty world of atoms. In a particularly picturesque passage that he had written somewhat earlier, he described sitting down to write out his lectures at two tables:

> Two tables! Yes; there are duplicates of every object about me. . . .
> One of them has been familiar to me from earliest years. It is a

commonplace object of that environment which I call the world. . . . Table No. 2 is my scientific table. It is a more recent acquaintance and I do not feel so familiar with it. . . . My scientific table is mostly emptiness. Sparsely scattered in that emptiness are numerous electric charges rushing about with great speed; but their combined bulk amounts to less than a billionth of the bulk of the table itself. . . .[11]

Eddington's unseen world was in some way the conceptual world of explanation built up by the physicist in trying to interpret natural phenomena. Jacob Bronowski put it well and even more forcefully in another context:

When we step through the gateway of the atom, we are in a world which our senses cannot experience. There is a new architecture there, a way that things are put together which we cannot know; we only try to picture it by analogy, a new act of the imagination. The architectural images come from the concrete world of our senses, because that is the only world that words describe. But all our ways of picturing the invisible are metaphors, likenesses that we snatch from the larger world of eye and ear and touch.[12]

While it is easy to accept Bronowski's lyrical description of the unseen world of the small, we normally neglect the fact that the world of the very large and very old is also a kind of architectural reconstruction, inferred from blackened grains on photographic plates and counts of electrons on ingenious photosensitive detectors. The intimidating vastness of time and space is of our own incredible construction. As we become more sophisticated in constructing our view of the cosmos, we must likewise become more nuanced in our view of God, of the divine. It makes no sense to drive an oxcart down the interstate.

Eddington, in his 1929 lecture, was grappling with something still more profound than the metaphysics of an invisible atomic world. He was bringing to the fore epistemology itself, as becomes clear in the following poignant passage:

It seems right at this point to say a few words in relation to the question of a Personal God. I suppose every serious thinker is rather afraid of this term which might seem to imply he pictures deity on a throne in the sky after the manner of medieval painters. There is a tendency to substitute such terms as "omnipotent force" or even a "fourth dimension." If the idea is merely to find a wording which shall be sufficiently vague, it is somewhat unsuitable for the scientist to whom the words "force" and "dimension" convey something entirely precise and defined. On the other hand, my impression of psychology suggests that the word "person" might be considered vague enough as it stands. But leaving aside verbal questions, I believe that the thought that lies behind this reaction is unsound. It is, I think, of the very essence of the unseen world that the conception of personality should dominate it. Force, energy, dimensions belong to the world of symbols; it is out of such conceptions that we have built up the external world of physics. What other conceptions have we? . . . We have to build the spiritual world out of symbols taken from our own personality, as we build the scientific world out of the symbols of the mathematician.[13]

When we ask how large is God? we are no longer asking about the scale of heaven or the physical nearness of the deity. We are necessarily using the vocabulary not just of everyday discourse but of science, philosophy, and theology, to build a coherent conception of a mystery so much greater than our own minds can encompass that we can only partially and incompletely come to terms with it. This is not an easy task. Our knowledge of the divine does not come from neon signs in the sky, from a message on the Internet magically created out of static, or from a set of scientific experiments. It must come from human channels, via human thought.

Elijah, fearing for his life and hiding in a cave in the wilderness, had a vision to go and stand upon the mount. "And behold, the Lord passed by, and a great and strong wind rent the mountains, and broke in pieces the rocks before the Lord, but the Lord was not in the wind, and after the wind an earthquake, but the Lord was not in the earthquake, and after the earthquake a fire, but the Lord was not in the fire, and after the fire a still, small voice."[14] God speaks to us, and to his prophets, through our thoughts and our deepest reflections, with a still, small voice. Our view of God, our understanding of God, comes from thinking about God as we observe the universe, our fellow creatures, and the historical record. A worshipful, religious view of the cosmos is not necessarily any less true than a scientific world view is true. Just because these insights are humanly constructed does not mean the results are invalid. After all, our entire scientific edifice is also humanly constructed.

Astronomers and physicists are continually pressing forward in the search for scientific knowledge, not only investigating the details of the origin and evolution of stars and galaxies but even daring to ask if our observed universe is merely one section of a still larger cosmos. Similarly, theologians and prophets are continually exploring the frontiers of spiritual knowledge. As our concepts of the cosmos have grown, so should our spiritual knowledge increase. As scientists, we can perhaps be forgiven for an occasional arrogant triumphalism over our spectacular successes in understanding the vast size and age of the universe. But as theologians, we can only be humble before the vastness of what we can still hope to learn about God.

When the Psalmist asked, "What is Man, that Thou art mindful of him?" I would say that the statement reflects the overwhelming majesty of what a creating superintelligence must be, a very large God indeed. But I would also turn to another Scripture, Genesis 1:27, "God created man in his own image, male and female created he them," and I would answer that as contemplative beings created in the image of God with attributes of creativity, conscience, and self-consciousness, we are central to the purposes of the cosmos. Understanding the cosmos is part of that purpose. Understanding humankind's role also is part of that purpose. The Book of Nature *and* the Book of Scripture. For me, faith is not blind faith but trust. And I trust that as we learn more about the vast cosmic scope and our place within it, our sense of the spiritual world will never atrophy but will ever expand.

Notes

1. Genesis 11:1-4, abridged from *The New English Bible* (1970)
2. According to E.A. Speiser, former chairman of the department of oriental antiquities at the University of Pennsylvania, Apsu is, among other things, a poetic term for the boundless expanse of the sky. *The Anchor Bible: Genesis* (Garden City, N.Y.: Doubleday, 1964), p. 75.
3. When told by a fellow historian of science that according to Aristotle the goodness of God kept the celestial spheres in motion, I could hardly believe it, but he promptly found the reference for me in the *Metaphysics* XII, ch. 7.
4. Leonard Digges, *A Prognostication Everlasting* (London, 1576) fol. 43.
5. Christiaan Huygens, *The Celestial Worlds Discover'd* (London, 1698), pp. 154-55.
6. William Herschel, "Account of some Observations tending to investigate the Construction of the Heavens," *Philosophical Transactions,* 74 (1784), pp. 437-51, reprinted in *the Scientific Papers of Sir William Herschel* (London: Royal Society and Royal Astronomical Society, 1912), vol. I, p. 166.
7. William Herschel, "On the Existence of Space," *Scientific Papers of Sir William Herschel* (London: Royal Society and Royal Astronomical Society, 1912), vol. I, p. lxxxv.
8. From "The Great Explosion" in Robinson Jeffers, *The Beginning and the Ending and Other Poems* (New York: Random House, 1963).
9. Lord Kelvin, "On the Dissipation of Energy," *Fortnightly Review,* 57 (o.s.) (March 1892), 313-21. Quoted by Joe D. Burchfield in *Lord Kelvin and the Age of the Earth* (New York: Science History Publications, 1975), pp. 42-43.
10. Arthur S. Eve, *Rutherford* (Cambridge: Cambridge University Press, 1939), p. 107.
11. Arthur Eddington, *The Nature of the Physical World* (Cambridge: Cambridge University Press, 1928), pp. xi, xii, xiii.
12. Jacob Bronowski, *The Ascent of Man* (Boston: Little, Brown, 1973) p. 340.
13. Arthur Eddington, *Science and the Unseen World* (New York: Macmillan, 1929), pp. 81-82.
14. I Kings 19:11-12, *Revised Standard Version*.

The Two Windows

Freeman J. Dyson

Mutability of Science

When we ask the question, How large is God? we are asking whether God transcends all our concepts and images. Another way to ask the same question is, How small is human understanding? Human understanding of the natural world is the subject matter of science. Another way to ask how large is God? is to ask, "how small is science?" Science has achieved amazing success in understanding the details of the world that we see around us. The success of science has led many people to believe that science can explain everything. But it would be even more amazing if science had no limits to its power. Humans are, after all, a species of ape that only recently climbed down from the trees. All our understanding of nature is based on human language. And human language is a tool contingent on the particular history of our species. It would be amazing if human language could comprehend aspects of the universe that no human has seen or experienced. If there are minds in the universe larger than ours, it is unlikely that our language could encompass their thinking. It is unlikely that our science could explain their concepts.

We shall never know how small we are until we have searched the universe diligently for other forms of intelligent life. If we are lucky enough to discover alien creatures with radically different ways of interpreting the universe, then we shall have a clearer view of our own limits. It would be a shock to our pride, but it would not be surprising, if we found that the aliens had no mental process corresponding to our use of language. It would not be surprising if we found that they had nothing in their culture that we could recognize as science. Science may not have a meaning beyond our own species and our own culture.

The meaning of science changes from one century to another, even within our own culture. The voice of science in the culture of the nineteenth century was not the same as the voice of science today. When we listen to a voice of science from the nineteenth century, it already sounds alien and distant. A typical voice from that century is Alfred Smee, medical doctor, botanist, inventor, prolific writer, and active participant in the intellectual life of London. He held firm views concerning religion and science. For him, religion and science both were universally valid. In his book, *The Mind of Man,* published in 1875, he writes:

> The results of the true reason of man are identical with the laws of God, and the one originating inductively from the human mind should accord deductively with the results which are obtained by the ordinances of religion. . . . Wherever religion and science do not exactly accord, the discrepancy marks error. It is then worth any labour to make them agree, by the conjoined operations of the labourers in religion and science, that truth may prevail.[1]

Alfred Smee speaks eloquently for the nineteenth century. Much of what he said is still valid today. But the twentieth century has brought us broader views, both of religion and of science. The twentieth century no longer sees religion and science as Smee saw them, as a unique set of laws of God and a unique set of laws of physics. The twentieth century sees religion as a tapestry of traditions and beliefs, expressing in various ways the human experience of belonging to a spiritual universe. The twentieth century sees science as a web of ideas and disciplines, expressing in various ways our practical knowledge of the universe and of ourselves. Smee's vision is no longer ours. Religion and science are too diverse and too rapidly evolving.

Just as religion and science have changed and grown in the hundred years since Smee wrote, we must expect that they will continue to change and grow in the twenty-first century. The most that we should ask of religion and science in the future is that they should respect each other's autonomy and should cooperate in tackling ethical problems. If the twenty-first century is to bring relief from wholesale misery and violence, it can only be by the active cooperation of religion and science in the struggle for social justice and international peace. For harmonious cooperation on the political level to be possible, science and religion must live together with mutual respect on the intellectual level. Religion and science should not view each other as two systems of laws that must be forced into exact accord. A better metaphor to describe religion and science today is as two windows, looking out on the world in different directions. Some of us see more clearly through one window, some of us through the other. If we listen to each other's stories of what we see, we may gain a deeper understanding than we could reach through either window separately.

Science and religion give us views of the universe that are both illuminating and both, to some degree, true. But they cannot be seen simultaneously. They are an example of the situation that Niels Bohr called complementarity, a situation that occurs in many areas of science and human life, when a single point of view cannot give a complete description of things. The most famous example of complementarity is the first thing God created, light. When one looks at light in one way it is a wave, and when one looks at it another way it is a particle, but one can never see the wave and the particle at the same time. The idea of complementarity explains why all our views of ourselves and of the universe are incomplete. It adds enormously to the depth and mystery of things. The world is far more alive when we look at it through two windows instead of one.

Why are we unable to look through both windows simultaneously? Because the rules of the two games are different. The essence of religion is faith and the essence of science is doubt. Not all theologians subscribe to any particular faith, but theology without faith would have no meaning. People must believe in something before they can embark on theological inquiry. On the other hand, the essence of science was stated most clearly in a conversation with my friend and hero Richard Feynman, "Doubt is not an obstacle to understanding, doubt is the essence of understanding." People must doubt everything before they can embark on scientific enquiry.

Another reason why one cannot look through the two windows simultaneously is that they require different ways of looking. To look through the religion window, one has to be quiet. One has to meditate or pray or think or listen or read or write, opening the mind or soul to the still small voice that one hopes to hear. To look through the science window, one need only learn to handle a few technical tools and then hammer away to one's heart's content. Science is gregarious and noisy. It is mostly done by groups rather than by individuals. The joy of doing good science is like the joy of helping to put a solid roof on a house. Science has more in common with house-building than with philosophy.

Since I like to deal with particular instances rather than with abstract generalities, I will look in detail at three examples of things lying on the border between science and religion, things that can be glimpsed in different ways through both windows. The first example is extrasolar planets, the second is extraterrestrial intelligent creatures, the third is extrasensory perception. The planets lie unambiguously within the territory of science, the alien intelligences lie precariously close to the border, and extrasensory perception lies clearly outside. These examples are chosen to illustrate in a practical way both the power and the limitations of science. At the end, I will come back to the general question, whether we see the world more clearly through one window or through two. Einstein, the great unifier, saw more clearly through a single window. Einstein saw the world as Alfred Smee saw it, as the sensorium of God, ruled by inflexible laws. I will explain why Einstein was an exception, and why the majority of scientists find his single view too narrow to encompass the unfolding prodigality of nature.

Plurality of Worlds

One of the great questions that can be examined either through the window of science or through the window of religion is the existence of planets orbiting around other stars beyond the solar system. Since our sun appears to be a typical star, it seems reasonable to expect that other stars will be accompanied by families of planets. For 60 years, beginning in the 1930s, astronomers searched in vain for evidence of planets around alien stars. Now suddenly, in the past five years, 12 extrasolar planets have been found. This happened because astronomers acquired new tools that increased the precision of their observations. The first family of extrasolar planets was discovered in 1992 by Alexander Wolszczan, a young radioastronomer who teaches at Pennsylvania State University. The news of his discovery was greeted by the community of astronomers in Princeton with considerable skepticism. I was lucky to be present when Wolszczan came to Princeton to confront the skeptics. This was a historic occasion. It provides an excellent example of the way we look at important discoveries through the window of science.

Alexander Wolszczan sat down to lunch with about fifty astronomers from Princeton and neighboring communities. The proceedings were entirely informal and friendly, but there was high tension in the air. Wolszczan had claimed to find planets orbiting around the wrong kind of star, not a normal star like the Sun but a tiny rapidly spinning object known to astronomers as a millisecond pulsar. A millisecond pulsar is as different from our Sun as it is possible to be. It is a collapsed remnant, only a few miles in diameter, that is formed when a giant star explodes at the end of its life. The explosion drives off into space the massive envelope of the star, leaving behind only the tiny collapsed core. If the giant star had possessed a family of planets before the explosion, it was extremely unlikely that any of them would have survived and remained attached to the remnant. It was equally difficult to imagine new planets being formed out of the debris from the explosion, since the debris is expelled from the star with enormous velocity. The prevailing view among astronomers was that a millisecond pulsar was the least likely type of star to be accompanied by planets. Wolszczan himself shared this view. The last thing he expected to find when he observed his pulsar was planets. He was not searching for planets. He came to believe that he had found planets only after all other explanations of his observations had failed.

The evidence for Wolszczan's planets was highly indirect. The only thing that he measured directly was the timing of the pulses of radio waves emitted by the pulsar as it rotates several hundred times per second. He measured the timing of the pulses with extraordinary accuracy, made possible by new tools, new clocks, and new computer programs. The pulsar itself is a clock, its pulses clicking with a regularity comparable to the most accurate man-made clocks. But Wolszczan, comparing the pulsar clock with his atomic clocks to microsecond accuracy over several years, found strange discrepancies. The timing of the pulses wandered, sometimes speeding up and sometimes slowing down, in a way that could be explained as the disturbing effect of two planets orbiting around the pulsar. From the detailed pattern of the wandering, Wolszczan could deduce the orbits of the planets and could calculate that they each must weigh about three times as much as the Earth.

The first thing that a good scientist does when confronted with an important discovery is to try to prove it wrong. Wolszczan, before telling anybody else about it, tried very hard to prove himself wrong. He checked and rechecked his apparatus and his computer programs; he searched for possible sources of interfering radio signals; he examined every instrumental effect that might have misled him. Only after all attempts to prove himself wrong had failed did he announce his discovery. As a result, when he came to Princeton to have lunch with the assembled astronomers, he was well prepared. Each astronomer who doubted the reality of Wolszczan's planets took a turn as prosecuting attorney, asking sharp questions about the details of the observations and trying to find weak points in Wolszczan's analysis. Wolszczan answered each question calmly and completely, showing that he had asked himself the same question and answered it long before. At the end of the lunch, there were no more questions. Wolszczan came through the ordeal victorious, and the skeptics gave him a friendly ovation. That is the way we see things through the window of science: Wolszczan's planets proclaiming their existence against all theoretical expectations; Wolszczan's skill and integrity overcoming all opposition from his more senior colleagues.

During the four years since Wolszczan's discovery, ten planets have been found by other astronomers. In astronomy as in other areas of science, it is always easier to discover additional examples of a new type of object after the first one is found. The astronomers who made the later discoveries had less skepticism to overcome than Wolszczan. The later discoveries differ from Wolszczan's in two other ways. First, the planets belong to ordinary stars like the Sun and not to millisecond pulsars. Second, the planets have much larger masses, several hundred times Earth's mass, unlike Wolszczan's planets with only three times Earth's mass. The newer planets are giants like Jupiter rather than dwarfs like our Earth. It was inevitable that the newer planets would have big masses. For a planet around a normal star to be detectable, it must have a mass comparable with Jupiter.

The most interesting planets from a human perspective, those that have a mass like Earth and belong to a star like the Sun, are undetectable with existing instruments. Planets with Earth-like mass can be detected at present only if they belong to millisecond pulsars. It is no accident that the only planets found until now with Earth-like mass are those discovered by Wolszczan. Almost all astronomers believe, after the recent discoveries, that planets with Earth-like mass orbiting stars like the Sun are abundant in the universe. But such planets will not be detected until new instruments or new techniques of observation are developed.

The plurality of worlds has been a central theme of religious inquiry, long before it became accessible to scientific observation. In the sixteenth century, Giordano Bruno argued that the greatness of God was better demonstrated in the creation of many worlds than in the creation of only one. Bruno preached this doctrine all over Europe, to Catholics and Protestants, and found in many places a favorable response. Only when he made the mistake of returning to his native Italy did he run into serious trouble. He was dragged to Rome, condemned by the Inquisition as a heretic, and burned at the stake. But his death did not put an end to his doctrines.

Gradually, as the Catholic and Protestant churches became more tolerant, the plurality of worlds ceased to be a mortal heresy and became an acceptable subject for theological speculation. In the eighteenth century, when the immense scale of the astronomical universe was revealed by William Herschel's telescopes, liberal churchmen were not slow to claim the millions of worlds in the sky as evidence of God's greatness and to imagine the millions of stars to be accompanied by millions of families of planets. In the twentieth century, belief in the plurality of worlds has become almost orthodox. The recent discoveries, confirming the notion that planets are abundant in the universe, came as no surprise to religious believers.

Plurality of Intelligent Species

Another question that lies on the boundary between science and religion is the existence of intelligent creatures elsewhere in the universe. Unlike extrasolar planets, extrasolar intelligent creatures have not been found. But the search for intelligent creatures continues, like the search for planets, as a serious program conducted with the highest standards of scientific rigor. If ever an object is found in the sky with behavior indicating the presence of intelligence, the evidence will be scrutinized even more thoroughly than the evidence for Wolszczan's planets.

One of the serious searches for extraterrestrial intelligence is directed by Paul Horowitz at Harvard University. Horowitz, like Wolszczan, is using the tools of radioastronomy for his search. He is looking for precisely tuned radio signals of a kind that no known astronomical object can produce. His main problem is to make sure that all signals produced by human activities on Earth or in space are excluded from his system of detectors. Horowitz is wisely more afraid of misidentifying a human signal as alien than of misidentifying an alien signal as human. He will not claim any signal to be alien until he has tried in all possible ways to explain it as a result of instrumental error or human interference.

Many learned papers have been published attempting to calculate the probability that extraterrestrial intelligent creatures may exist or the probability that they may communicate with each other by radio. The purpose of such calculations is to guide the search programs in the belief that we would have a better chance of finding intelligent aliens if we could guess in advance their probable distribution in the universe and their probable behavior. The example of Wolszczan's planets shows that such calculations of probability are worthless. Wolszczan found his planets in a place that all the experts believed to be wildly improbable. If his search had been guided by calculations of probability, he would never have found them. The most useful criterion to guide a search program is not probability but detectability. Wolszczan was the first to discover alien planets, because planets orbiting around a millisecond pulsar are detectable, whereas similar planets orbiting around a normal star are at present undetectable.

When we search for extraterrestrial intelligence, the same principle should guide us. We cannot hope to calculate the probability of any alien activity, but we can reliably calculate its detectability. Horowitz understands this logic. He concentrates his search on signals that are detectable, whether or not they may be judged to be probable. Again and again in the long history of astronomy, great discoveries were made because new types of objects became detectable when nobody would have imagined them to be probable.

The search for extraterrestrial intelligence is considered by many scientists to be dubious science. It is suspect for two reasons: First, it has been pursued for 35 years without a single positive result; second, it has been accompanied by a great deal of wishful thinking and exaggerated expectation. Although Horowitz' observational program is solidly based in science, it sits close to the edge where science ends and science fiction begins. Intelligent aliens always have been a major theme of science fiction. And some scientists find any association with science fiction distasteful.

For the majority of scientists, a serious interest in science fiction is entirely compatible with a serious interest in science. As long as we keep clearly in mind the difference between fact and fiction, science fiction can be a useful source of scientific ideas. My own involvement in the search for extraterrestrial intelligence resulted directly from my reading of the science fiction novel *Star Maker*, written by Olaf Stapledon in 1937. The novel is like *Gulliver's Travels*, an account of a series of visits to extraordinary places, mixed with philosophical meditations concerning the smallness of humanity and the greatness of the universe. Half way through his voyage of imagination, Stapledon describes a vision of a galaxy densely populated by alien intelligent civilizations. "Not only was every solar system now surrounded by a gauze of light traps, which focussed the escaping solar energy for intelligent use, so that the whole galaxy was dimmed, but many stars that were not suited to be suns were disintegrated, and rifled of their prodigious stores of sub-atomic energy."[2] This vision has remained vivid in my memory since I read *Star Maker* in 1945.

In the early 1960s, the first scientific search for extraterrestrial intelligence was begun by the radioastronomer Frank Drake. Drake, like Paul Horowitz in more recent times, was listening with a radiotelescope for finely tuned signals that might be evidence of alien communications. Searches of this kind, using radiotelescopes, could only be successful if the aliens were interested in communicating by radio. When I heard about Drake's search, I asked myself, What can we do to find aliens who are not interested in communication? and I recalled Stapledon's vision of a densely populated galaxy. If any part of our own galaxy or of a neighboring galaxy were populated with aliens as densely as the galaxy envisaged by Stapledon, the aliens would be easily detectable, even if they were not communicating by radio. The enormous structures built by the aliens to house themselves and their machines, the . . . "gauze of light traps, which focussed the escaping energy for intelligent use, so that the whole galaxy was dimmed . . ." would be detectable as sources of heat radiation emitted from their warm surfaces. A search with infrared telescopes, on the ground or in space, would reveal such artificial sources of heat radiation if they exist.

I suggested that a search with infrared telescopes for noncommunicating aliens should be begun, complementing the search for communicating aliens with radiotelescopes. The infrared search would be based, like the radio search, on the principle that we should search for what is detectable, whether or not we believe it to be probable. We have no reliable way to estimate the probability that our galaxy should contain a dense population of aliens. All that we can say with certainty is that a sufficiently dense population will be detectable. Until now, no systematic infrared search has been undertaken. Vast populations of intelligent aliens may exist as Stapledon imagined them, waiting for our infrared telescopes to detect them. I will not be surprised if the first discovery of extraterrestrial intelligence is made, like the discovery of Wolszczan's planets, by an astronomer who searches the sky for expected kinds of natural objects and stumbles upon something totally unexpected.

Many of the best writers of science fiction are religious believers, and some of them are uninhibited in mixing religion into their fiction. Science fiction may, on occasion, serve as a bridge between religion and science. Olaf Stapledon's *Star Maker* is a fine example of such a bridge. It is a bold attempt to create a theological framework for a history of the universe. Stapledon's God is an experimenter, creating a succession of universes, learning from his failures, and building on his successes. Our universe is neither his first nor his last. It is a flawed experiment, destined to end in tragedy, but providing a basis for other universes with fewer flaws. Stapledon agrees with the poet William Blake, who wrote: "To be an error and to be cast out is a part of God's design[3]." Stapledon answers the question, How large is God? with a succession of revelations of ever-increasing scale, beginning with a single planet, then moving through systems of stars and galaxies, then encompassing our entire universe, and finally going beyond our universe to other universes inconceivable to human minds.

After creating our universe, according to Stapledon:

> The Star Maker contemplated his work. And he saw that it was good. . . . He knew that this creature, though imperfect, though a mere creature, a mere figment of his own creative power, was yet in a manner more real than himself. As he discriminated its virtue and its weakness, his own perception and his own skill matured. . . Thus, little by little it came about, as so often before, that the Star Maker outgrew his creature. Increasingly he frowned upon the loveliness that he still cherished. Then, seemingly with a conflict of reverence and impatience, he set our cosmos in its place among his other works.[4]

At the end of *Star Maker*, we have come very far beyond the borders of science. Stapledon is no longer writing science fiction. He has moved into a new genre of fiction that we might call theofiction, and looks at his characters through the window of religion rather than through the window of science. He is not alone in the new genre. Other first-rate writers, notably Clive Lewis and Madeleine L'Engle, have given us their different imaginative visions of a spiritual world.

Extrasensory Perception

Further from the mainstream of science than extrasolar planets and extraterrestrial intelligence lies the domain of extrasensory perception. The belief that certain individuals at certain times have the ability to see and know things outside the normal channels of perception is deeply embedded in many human cultures. Many religions have incorporated this belief into ritual and dogma. Classical Greek religion had its oracles, ancient Judaism had its prophets and seers, Christianity has its saints and miracles.

Isaac Newton, the scientist who first demonstrated mathematically the working of the physical universe, nevertheless believed in angels and occult powers. Newton's mind was razor-sharp, whether he was dealing with the geometry of planetary orbits or with the interpretation of scripture. Newton showed a thoroughly modern skepticism in dismissing the evidence of paranormal phenomena reported by certain ascetic monks:

> I find it was general complaint among them that upon their entring into the profession of a Monastick life they found themselves more tempted in the flesh than before, and those who became stricter professors thereof and on that account went by degrees further into the wilderness than others did, complained most of all of temptations. The reason they gave of it was that the devil tempted them most who were most enemies and fought most against him: but the true reason was partly that the desire was inflamed by prohibition of lawful marriage, and partly that the profession of chastity and daily fasting on that account put them perpetually in mind of what they strove against, and their idle lives gave liberty to their thoughts to follow their inclinations. The way to chastity is not to struggle with incontinent thoughts but to avert the thoughts by some imployments, or by reading, or by meditating on other things, or by convers. By immoderate fasting the body is also put out of its due temper and for want of sleep the fansy is invigorated about whatever it sets itself upon, and by degrees inclines towards a delirium, in so much that those Monks who fasted most arrived to a state of seeing ap-

paritions of women and other shapes and of hearing their voices in such a lively manner as made them often think the visions true apparitions of the Devil tempting them to lust. Thus while we pray that God would not lead us into temptation these men ran themselves headlong into it.[5]

Newton was as aware as any modern psychologist of the unreliability of human perceptions. He was also open-minded about the reality of supernatural events reported in the scriptures. In an unpublished commentary on the biblical book of Revelation, Newton writes:

> If you ask where this heavenly city is, I answer, I do not know. It becomes not a blind man to talk of colours. Further than I am informed by the prophesies I know nothing. But this I say, that as fishes in water ascend and descend, move whether they will and rest where they will, so may Angels and Christ and the Children of the resurrection do in the air and heavens. 'Tis not the place but the state which makes heaven and happiness.[6]

He found no inconsistency between a belief in the scriptures and a sharply critical view of their meaning.

During the past two centuries, many distinguished scientists have attempted to investigate the possibilities of extrasensory perception among their contemporaries. Two distinct methods of investigation have been pursued: the anecdotal and the scientific. The pioneers of the anecdotal method belonged to the Society for Psychical Research, founded by Frederick Myers in London in 1882. When instances of extrasensory perception were reported, they collected evidence with great care, examined witnesses, and tried to obtain independent corroboration and documentation of their statements. The findings and the supporting documents were published in the *Journal of the Society for Psychical Research,* always in a sober and professional style. Many of the case histories in the *Journal* are well supported by first-hand evidence and contemporary documents. Many of the stories are attested by well-known people of high reputation. An impartial reader cannot fail to be impressed by the seriousness of the investigators. Any reader not committed in advance to dogmatic belief in the nonexistence of extrasensory perception must allow the possibility that some of the stories may be true. On the other hand, a reader not committed in advance to belief in the reality of extrasensory perception must allow the possibility that the stories are untrue. In no case can the possibilities of fraud and self-deception be rigorously excluded.

Anecdotal evidence can never add up to scientific proof. The essence of scientific proof is repeatability. The reality of a phenomenon is scientifically proved only if it can be repeated and observed under controlled conditions. The reported instances of extrasensory perception can never be repeated under conditions controlled by a skeptical inquirer. The anecdotal evidence, no matter how carefully it may be documented, cannot be tested by the methods of science.

After the stories collected by the pioneers of psychical research had failed to convince the skeptics, a later group of researchers attacked the problem of extrasensory perception with controlled experiments. The typical controlled experiment was done with cards carrying five different designs in equal numbers. An experiment to test telepathy would be done with one person, the "sender," sitting alone in a closed room and looking at a succession of cards, while a second person, the "receiver," sitting alone in another room, would record a succession of guesses, guessing the design on each card as the sender looked at it. The card-guessing would be scored by counting the number of correct guesses or "hits." The game was played with many variations to test different modes of extrasensory perception and with many safeguards to prevent covert signals from passing between the participants and the experimenter.

The results of the game were often encouraging at the beginning but usually disappointing at the end. The receiver would sometimes begin a session with a long series of guesses containing substantially more hits than the one out of five expected from random chance. After the first day, results would be tabulated and hopes would rise high. But then, as the game went on from day to day, the fraction of hits would decline until in the end it did not differ significantly from one-fifth. Great efforts were made, and the patience of participants strained to the breaking point, with no useful result. When occasionally an experiment was reported with a startlingly high fraction of hits sustained for a long period, further investigation of the experimental procedures always revealed that precaution against covert communications had been neglected. No well-controlled experiment ever gave convincingly positive results. No initial successes could be consistently repeated.

The failure of extrasensory perception experiments can be explained in two ways. First, extrasensory perception may not exist. This is the simpler explanation and is accepted by almost all scientists, who are not troubled by the anecdotal evidence in favor of extrasensory perception. They are glad to dismiss the anecdotal evidence as scientifically worthless. But a second explanation is possible, allowing us to give some weight to the positive anecdotal evidence while accepting the negative scientific evidence. The second explanation is that extrasensory perception may be real but works in ways that make it incompatible with scientific investigation.

According to the second explanation, there is a relation of complementarity between extrasensory perception and controlled experiment, similar to the relation of complementarity between the wave nature and the particle nature of light. One can observe light behaving as a wave and one can observe light behaving as a collection of particles, but one cannot observe both behaviors simultaneously. The apparatus required to observe light as particles causes the wave-like properties of light to disappear. In the same way, according to the second explanation, extrasensory perception can be observed in the uncontrolled experiences of human life, but the severe constraints of any controlled experimental situation make it disappear.

Anyone who has taken part in a card-guessing experiment readily understands how the conditions of the experiment can destroy the more delicate human faculties. The experiment is unbearably tedious, requiring the participants to spend long hours carrying out trivial and repetitive tasks. The anecdotal evidence indicates that extrasensory perception, if it occurs at all, is associated with strong emotions and important events. The tedium and triviality of the experiments seem expressly designed to exclude the intense feelings that extrasensory perception may require.

It is easy to imagine that the eagerness of the participants at the beginning of an experiment might allow some slight effect of extrasensory perception to be recorded, and then, as eagerness is overcome by boredom, the effect fades into significance. Nobody has found a way to design an experiment that is both rigorously controlled and emotionally interesting to the participants. It may well be that such an experiment is impossible, and the requirements of emotional intensity and scientific control are necessarily incompatible.

If extrasensory perception is real and the second explanation of the failure to measure it scientifically is valid, then it is a purely mental faculty like love and hate and the appreciation of natural beauty, not a physical faculty like sight and hearing. It would be as unreasonable to attempt a scientific measurement of the act of extrasensory perception as to attempt a scientific measurement of the act of falling in love. Many of the most essential human faculties are inaccessible to scientific measurement.

Extrasensory perception is only one example of many dubious phenomena that may or may not exist in our universe. Other examples are angels, ghosts, devils, divine inspirations, and miracles. These phenomena are all beyond the reach of science and belong to the domain of religion. Most of my scientist friends believe dogmatically that such phenomena do not exist. I prefer, following the good example of Isaac Newton, to leave all possibilities open. As a scientist, I am not compelled to believe dogmatically in their nonexistence, and as a religious person, I am not compelled to believe dogmatically in their existence. I look out at the universe through the two windows of science and religion. Through the science window, I see only a small fragment of a huge landscape that we have barely begun to explore, full of puzzles and mysteries. Through the religion window, I see fragmentary glimpses of an even wider landscape inaccessible to the tools of science. Nothing could be more absurd than the belief that either of the two views, or both of them together, come close to being complete or final.

Einstein's View of Science

Einstein was a rare kind of scientist, an exception to all rules. Einstein would not have agreed with my picture of science and religion as two windows looking in different directions. Einstein's personal religion was inseparable from his science. For him there was only a single window. I disagree with him because I am an ordinary scientist and he was not.

I give the name science-worship to the faith that Einstein professes in his later writings. It is a seductive ideology because it gives to science a special place, above and beyond other human activities. It gives to scientists the moral authority of a priesthood. And Einstein was the great protagonist of science-worship. The essence of the ideology is contained in his famous statement (made at a symposium on science and religion in 1941 when he was sixty-two years old):

> Science can only be created by those who are thoroughly imbued with the aspiration toward truth and understanding. This source of feeling, however, springs from the sphere of religion. To this there also belongs the faith in the possibility that the regulations valid for the world of existence are rational, that is, comprehensible to reason. I cannot conceive of a genuine scientist without that profound faith. The situation may be expressed by an image: science without religion is lame, religion without science is blind.[7]

Einstein did not subscribe to the beliefs of any conventional religion. He used the word "religion" to mean his own private faith in the rationality of nature. But his statement has the flavor of ideological conviction. It says, in effect, anyone who does not share my faith is not a genuine scientist.

It is possible to have the greatest possible respect for Einstein as a scientist and as a human being, and still disagree with his ideology. It is quite untrue that science without religion is lame. Many excellent scientists have no profound faith of any kind. Einstein's statement excludes from genuine science the great multitude of people for whom science is a way to earn a living rather than a way to contemplate ultimate reality. He could not conceive of a genuine scientist without profound faith. This meant that he took no interest in the work of the majority of scientists who did not share his exalted view of science. At the end of his life, he was left isolated because science forged ahead with fruitful discoveries that did not fit into his narrow view of what science ought to be.

The last ten years of Einstein's life were years of extraordinarily rapid progress in science. Young people in Italy, England, and the United States were discovering new and unexpected species of elementary particles. The new technology of microwave electronics allowed experimenters to explore the behavior of atoms with unprecedented accuracy. Every month, new experiments and new theories were announced. But for Einstein, all this joyful activity was not genuine science.

He was out of touch because he could not broaden his interests to keep pace with the broadening scope of science. His single window gave him a narrow view. It has always seemed to me that his narrowness resulted from making science itself the object of religious worship. He tried to identify science with religion, to bring science and religion so close together that they could not be separated. This religion of science-worship was right for him personally, but it is wrong for the majority of scientists and for the majority of religious believers. It denies both to science and to religion the freedom to be themselves, the freedom to be different.

I have seen the same tragedy played out in the lives of many scientists and teachers. A too intense science-worship disqualifies them from doing useful work in the real world. Science-worship is often accompanied by an intellectual snobbery that leads them to despise practical applications of science. They have Einstein's ideology without his genius, a combination that condemns them to lives of frustration. My scientific friends who clamored so passionately for the building of the superconducting supercollider were also victims of science-worship. They could not understand that their pleading for this project that seemed to them so noble and holy was seen by others as an exhibition of normal human greed.

Einstein was wrong in ascribing holiness to science. The holiness is in the universe that the science is trying to explain. The science is a makeshift sketch that we shall replace as soon as we have explored a little further. God is in the richness of the phenomena, not in the details of the science. The phenomena are God's business, the science is ours. Science-worship is a new form of an old idolatry, putting a man-made image in the place of God. The essential error of Einstein was to say that science is holy, just as the essential error of the creationists is to say that religion is scientific. In truth, the glory of science lies in the clever use of tools, and the glory of religion lies in the poetry of worship. The two glories may co-exist in the same human soul, but they are not the same.

Notes

1. Alfred Smee, *The Mind of Man: Being a Natural System of Mental Philosophy* (London: George Bell and Sons, 1875). For this reference, I am indebted to George B. Dyson.
2. Olaf Stapledon, *Star Maker* (1937; reprint, Los Angeles: Tarcher, 1987) p. 179.
3. William Blake, "A Vision of the Last Judgment," Rossetti Manuscript, 1810, in *Poetry and Prose of William Blake,* ed. Geoffrey Keynes (London: Nonesuch Press, 1939) p. 647.
4. Stapledon, *Star Maker*, pp. 241–242 (Tarcher).
5. Frank E. Manuel, *The Religion of Isaac Newton* (Oxford: Oxford University Press, 1974) pp. 13.
6. *Ibid.,* p. 101.
7. Alfred Einstein, *Ideas and Opinions by Albert Einstein* (New York: Bonanza Books, 1984) p. 46.

PART 2

Approaches to Scientific and Theological Answers

Approaching God Through Paradox

F. Russell Stannard

It is difficult to think of science without also associating the idea of progress with it. Science moves forward, pushing back the frontiers, and embracing new theories in the light of fresh data gained. Scientific knowledge is always improving; its understanding of the workings of nature forever deepening. The modern scientific world view is clearly an improvement on that of past ages; we are convinced that we are closer to the truth now than at any previous time.

When it comes to religion, we appear to be dealing with a different kind of activity. It is commonly thought that, unlike science, religion is firmly rooted in the past. It reveres the teachings of the great men of long ago: Moses, Jesus, Muhammad, and the Buddha. Their words are enshrined in sacred scriptures that are not to be questioned. The truths of religion are fixed and unchanging.

But is that a fair assessment? Is religion so very different from science?

At the outset, one has to admit that it is easy to see how this common perception arises. The great religious figures of the past had extraordinary spiritual insights. In particular, if one believes, as a Christian, that Jesus was uniquely the Son of God, then his life and teachings constituted a special revelation of God that outranks all others—one against which the veracity of any future claims to spiritual understanding have to be tested. In science, there is no such anchor point in the past.

But even so, it has to be remembered that Jesus himself spoke of the coming of the spirit after his ascension—the spirit that would lead us into *all* truth. One interpretation is that there was yet more to learn about God and his ways—further truths to be revealed.

It is my contention that just as science makes progress so also does religion, or more strictly speaking, theology. In the same way as one would not wish to trade our modern scientific view of the world for

69

that of past generations, so too our understanding of God is an improvement on that of the past—it is deeper now than at any other time. And just as it is confidently expected that scientific advances will continue into the future, so there is still much to learn about God.

This is true not only of the *collective* understanding of the natural world and of its Creator but also of our own *individual* understanding of these matters. Throughout our own personal lives, we should be constantly replacing childish misconceptions about the world in which we live by more adequate notions based on the findings of science, and our understanding of God also ought to be in a continual state of revision and updating in the light of new experience and insight.

God of the Bible

To convince oneself that knowledge of God *does* change with time, one has only to rearrange the writings of the Bible in chronological order. Deciding on the correct order in which they were written can be problematic, but when this is done to the best of present knowledge, the following broad picture emerges.

We begin with the commonly held belief that there were many gods (or baals). Each was in charge of a given territory and its people. One of them, Yahweh, was unusual in that although he had his territory—a mountain in the Sinai desert—he *adopted* his people, the Jews. At that time, he had no interest in other people; he thought nothing of killing off the first-born of the Egyptians and subsequently drowning its army. He was a jealous, war-like god who was to leave his mountain and go with his people on their wanderings through the desert. He fought for them against the Canaanites.

But then it was discovered that Yahweh was more than a fighting god. As Elijah showed, one could pray to him for rain; he could be entrusted with the growth of crops in peacetime; it was not necessary to invoke the help of the local baals for this purpose. Through Hosea (learning from his relationship to his wayward wife), Yahweh was found to be a god of love: He was able to forgive his people when they went astray. Then the poor shepherd, Amos, emphasized that Yahweh was a god of justice—justice for the poor and downtrodden.

He was concerned about the plight of *individuals*, not just *nations*. Isaiah, accustomed as he was to life in the king's palace and the ways of the nobility, had revealed to him something of the splendour and power of Yahweh—his glory filling the *whole* earth and not just Judah. Yahweh ruled the whole world—the most powerful god of all.

During the desolation of the exile, when the leading Jews were deported to Babylon from Jerusalem and the great temple (the home of their god) was destroyed, Jeremiah taught that Yahweh was not confined to any temple but dwelt in the heart of the individual person, and for that reason, he would be with them wherever they went.

Then came the greatest insight of all. As recorded by Second Isaiah, Yahweh said:

"I am the Lord, and there is no other; apart from me, there is no other god."

He was the only God; there had never been any others. Recognition that he was the ruler of the entire world then made the way clear for him to be seen as the great Creator-God of the world and of all living creatures, including ourselves. Depending for our very existence on him, we all owe him our loyalty.

Finally, with the coming of Christ, God enters into his creation, taking on a human life. He demonstrates in the clearest possible way his limitless love for us in the manner of his suffering and death. And then through the Resurrection of Jesus, we have the evidence of his power over death and of the prospect of eternal life with him.

Such an account of the development of the idea of God through the ages, although sketchy and oversimplified, nevertheless makes clear that theology is anything but static and fixed for all time. Indeed, a case could be made that the cumulative improvement in the understanding of God that took place during biblical times was as far-reaching and dramatic in its way as any scientific advance in our own times.

God beyond Appearances

How is one able to make these advances in knowledge, whether we are thinking of knowledge of the world or of God? Essentially, it is a matter of going beyond superficial appearances and the seemingly "common-sense" notions of what is happening. First impressions are nearly always misleading. Instead, one needs careful observation. The information from these experiences has then to be organized systematically and interpreted.

In the field of science, for example, a seemingly obvious flat Earth, together with the special direction in nature called "down," had to be replaced by a round Earth and a law of gravity such that each particle of matter attracts every other particle of matter in whatever direction it might lie. It was found that, contrary to appearances, moving objects did *not* of their own nature slow down; if left to their own devices (i.e., in the absence of friction), they will continue to move at constant speed. Although energy appears to get "used up," nevertheless, it is conserved.

The bewildering array of hundreds of thousands of chemical substances, each with its own characteristic properties, can be made from just 92 different naturally occurring elements, and these, in turn can be described in terms of just three particles: proton, neutron, and electron. The multitude of seemingly capricious patterns of motion and behavior we see can all be accounted for in terms of adherence to just a handful of dynamical laws.

In a similar way, the changing understanding of God manifested in the Bible represented a movement away from first impressions and seemingly obvious solutions to more sophisticated observation and interpretation. For example, with the nations of the world constantly waging war against each other in the pursuit of conflicting goals and aspirations, it was not easy to recognize that there was but the one God to whom all sides owed equal allegiance. Similarly, when so much evil was seen to be arrayed against the good, it took time, reflection, and deep insight to distill from this the idea of God who is wholly good and loving.

Again, in the face of manifold injustice in the world, it was hard to discern the hand of a God of justice, until there was a firmer realization that there was more to life than our earthly existence. Finally, in a society where certain people—the rich and powerful—appeared so much more important and influential than others, it was not immediately obvious that in the eyes of God all individuals are of equal value.

Thus, it was only on mature reflection that the great prophets of the past, drawing upon their life experiences and circumstances, could collectively lead us from a picture of a divine world that was little more than a collection of petty, blood-thirsty warring tyrants to the vision of the one great Creator and ruler of the whole world—the loving heavenly father of all.

God the Father

So we reach the point of inquiring what further progress might be made in achieving an even deeper understanding of God. How might we speak of God in the twenty-first century?

It is difficult, if not impossible, to talk of God except through analogy and metaphor. Somehow we need concrete images of one sort or another, these being drawn from firsthand experience of life. The strongest metaphor for God to emerge from the Bible is that of the Father. But as with any metaphor, along with the positive, helpful, similarities that help to illuminate the unknown, there are those aspects that do not apply. As a prelude to seeing how our conception of God can be further enriched, we shall examine what the limitations might be of the "father" metaphor.

But, as a preliminary, perhaps we ought to address Freud's argument that the whole idea of a heavenly father figure is illusory.[1] Freud's view was that, having been dependent on the protection of our earthly father while young, on growing up we seek to prolong that sense of being looked after. A feature of the unconscious is that it can on occasion convince us that what we wish for has in fact come about. In this way, the unconscious, through the process of wish fulfilment, leads us to replace the earthly father with an imaginary father figure in Heaven.

There is much that one can say against Freud's suggestion. In the first place, one would have thought that dependence on the protection of a father was universal and that a divine father figure would be a universal feature of religions throughout the world, and for all times. But this is not so. Second, Freud's relationship with his own father was strained. Were one to apply the same type of psychoanalytic reasoning to him as he developed for the treatment of his patients, one would predict that his experiences would lead him to reject father figures—heavenly or otherwise. Atheism was something that Freud brought to his psychoanalytic theories; it did not arise out of them. And, indeed, Freud admitted that his findings regarding religion could have been given a different interpretation to the one he himself adopted.

That said, however, it would be dangerous and unwise to dismiss out of hand Freud's views on the matter. It has been shown that the type of heavenly father in which we believe can be strongly colored by the relationship one has had with one's own father: Whether he is regarded as a stern disciplinarian or one that is more easy going. One also suspects that the more relaxed attitude people have toward God today, particularly fear of future punishment in Hell, probably has much to do with the replacement of the heavy-handed Victorian stereotype of a father by the more democratic "let's talk-this-over-together-as-equals" attitude that the modern parent is expected to adopt toward children. To make the most of the "father" metaphor, one has to shake off the idiosyncratic connotations that this term carries for us, which arise out of the particularities of our own experience of an earthly father and also those passing associations that arise from currently accepted customs and conventions of society.

That done, there remains the problem of the specific sexual connotations associated with the metaphor. In a society dominated by males, it was only natural that the great religious leaders of the past were men rather than women. Had the Christ been female, for example, or had Jesus chosen women as apostles, it is doubtful they would have been accorded a hearing in the circumstances prevailing at that time. So, the all-male leadership of the early Church is no reason for believing that there is something inherently male about God.

In the changed social climate of today, where women are at last gaining the opportunity to shine in spheres of influence previously denied to them, the irrelevance of gender in most contexts is becoming increasingly clear. That being so, there seems little justification for thinking of God—the God of *all* people—as having a particular affinity with the 50 percent who happen to be male. Although there might remain some concern that the trend to accord to God's nature a feminine aspect might merely be a reflection of current trends in society (making God in society's own image), it seems to many that we have an example of the spirit of truth leading us not only toward the recognition that men and women have equal rights but that in God's very nature there is no preference for one sex over the other.

The idea of a heavenly parent who incorporates qualities drawn from both father and mother is obviously a more difficult conception to grasp than that of a straightforward father figure. We are being called upon to hold simultaneously in the mind somewhat contradictory metaphors, combining relevant features from both, while at the same time discarding the inappropriate aspects. But that is how we are likely to make further progress in understanding God. Once we recognize that in addition to all the positive aspects of the father metaphor (in terms of protection, discipline, training, guidance, wisdom, example, etc.), we can add those of the mother metaphor, we achieve an enriched conception.

For example, when it comes to creation, there is much to be said for the idea of the world springing from "the womb of God." It provides a picture that not only incorporates the dependence of the world on God but also the independent quality of the offspring once it has been given birth. The more usual metaphor in which God fashions the world as a potter working clay leads to a product that may or may not reflect closely its designer. It could be an anonymous pot that anyone could have made. But with the womb analogy, the child by its very genetic nature inherits much of the quality of the parent and carries within it the marks of its maker.

God the Creator

Having just then touched on God's role as Creator, we turn next to cosmology. Of all the branches of science, this is the one in which some of the most spectacular advances recently have been made. This new understanding of the origins of the universe was, of course, not available to people in biblical times, so we must ask whether this new information about the world has any additional light to shed on the character and role of its Creator.

It is currently accepted that the universe came into being through the big bang. Originally, all the contents of the vast universe we observe today were concentrated at a point. From this infinitesimal beginning, it expanded. The great swirling collections of stars we call galaxies are still receding from each other today in the aftermath of that great explosion. The cooled-down remnants of the fireball that accompanied that explosive event still bathe the Earth in cosmic microwave background radiation.

In his best-seller on cosmology, *A Brief History of Time*, Stephen Hawking, having described our modern understanding of the origins of the universe asks the question, What place then for a creator?[2] Is it the case that, far from providing additional information about God, modern cosmology has actually dispensed with a divine creator?

One reason for questioning the role of God in the big bang is the suggestion that it could have happened by itself; the world might have come into being spontaneously. To understand how this might have occurred, one needs to be acquainted with the probabilistic nature of quantum physics:

We are accustomed to the idea that to every effect there is a cause; the effect, then, in its turn, becomes a cause giving rise to the next effect down the causal chain. If one repeatedly sets up the same cause, under identical conditions, then the same effect will result. This deterministic behavior appears to fit in well with our everyday experience.

But appearances can be misleading. The fact that, strictly speaking, determinism is *not* the case, becomes obvious when dealing with very small objects such as individual subatomic particles. If one repeatedly sets up the identical initial state of the particles, one *cannot* predict with certainty what the effect will turn out to be. All one can do is assign various probabilities to a range of possible outcomes. One has simply to await what the particular outcome will be. This need to deal in probabilities rather than certainties is summed up in Werner Heisenberg's famous Uncertainty Principle.

Although this quantum uncertainty was discovered at the level of tiny subatomic particles, it underlies *all* behavior. The reason it is not obvious in everyday experience is that we are usually dealing with such large macroscopic objects that the quantum uncertainties become negligible compared with the quantities being measured; they, therefore, generally pass unnoticed.

This being so, what relevance has this for cosmology—the study of the universe that is the largest thing there is? The relevance becomes apparent when one recalls that the universe was not always big. Indeed, at the instant of the big bang, it presumably had no size at all. That, in turn, would mean that the uncertainties would have been infinite. It is this thought that prompts the idea that perhaps if one starts off with an initial state consisting of nothing, there might be a finite probability of a violent quantum fluctuation giving rise to a state of "something." This something begins infinitesimally small but promptly undergoes a big bang and develops into our familiar cosmos. For all we know, this sort of thing might be going on all the time. Universes might be spontaneously tumbling into existence all the time by these quantum fluctuations.

This, of course, is little more than speculation. Other universes by their very nature are not open to experimental verification from this universe. Also, we are talking of what supposedly happened at a point of infinite density, and there is no way our physics can handle a situation like that. Yet the spontaneous generation of universes, through the scope apparently afforded by quantum uncertainty, is an idea that appeals to many. So we must ask: Would such a possibility threaten the idea of a Creator-God?

It would certainly call into question an understanding of God that is based too heavily on the analogy of the potter fashioning a pot (or indeed the mother giving birth); this cosmic "pot" would spontaneously come into being without the need of any hands-on involvement from a cosmic "potter." Yet there would still remain a question: Why *quantum* physics? Why was quantum physics in charge of the process rather than some other form of physics? After all, it is quite easy to imagine worlds run on lines very different from those of our own (science fiction writers dream up fantastical alternative worlds all the time). Would it not require a God to decide which kind of laws would operate?

An idea has been advanced that perhaps the set of natural laws that operate in our world is the *only* set of laws there could be; alternatives are ruled out because, for some reason or other, they are inconsistent and would lead to contradictions. Whether there is any truth in this suggestion, one thing is clear: Such a claim could never be verified. The reason is that the laws lend themselves to mathematical expression. That being so, we can specify certain features that the laws (whatever they might be) *must* have because of the general constraints to which all mathematical structures conform.

For example, we know from general mathematical principles that, from within a mathematical structure, it is impossible to justify the axioms (in this case the fundamental physical laws) upon which the structure is based. We humans, being component parts of the universe, have representations that are part of the overall mathematical structure that governs the working of the universe; we ourselves are *within* the structure. That in turn means that we could not possibly justify why the mathematical structure (corresponding to the operation of the physical world) was based on that particular set of axioms (laws of nature) rather than some other. So the view that there could be one, and only one, set of laws namely those incorporating quantum uncertainty, is something that we could never prove; ultimately, it would have to be a matter of faith.

The alternative is to hold that there could be many different kinds of world. In this case, a choice *had* to be made as to which laws should apply. God's Creator role would then be rather different from the customary one. He would not be the light-the-fuse-and-retire type of god, but one who sets in place the laws. It is then the natural outworking of those laws that is responsible for our cosmos (and perhaps other worlds) coming into being.

Be that as it may, there is further reason for us to revise our understanding of what it means to regard God as the Creator:

The impression of the big bang is that it was an explosion much like any other explosion—only bigger. But that is not so. With an ordinary explosion, the bomb has an initial location. When it goes off, it flings out its debris to fill the rest of the surrounding space. With the big bang, however, there was *no* surrounding space. Not only was all matter and energy concentrated at a point but also all the space of our universe. It began infinitesimally small, then suddenly expanded. We can regard the continuing expansion of the universe today as a direct result of this process whereby the galaxies are being swept apart on a tide of expanding space. (The usual analogy is that of a sheet of rubber on which has been glued some coins. The coins represent the galaxies. As the sheet is pulled in all directions, the expanding rubber between the coins carries them apart.)

Thus, the big bang marked the creation not only of the contents of the universe but of space itself.

That in itself marks an important shift in our thinking. But there is more to follow. Since Einstein brought forward his theory of relativity, we have come to recognize that there is a very close connection between space and time. Although we experience them in totally different ways, measuring three-dimensional space with a ruler and one-dimensional time with a clock, nevertheless, appearances are (again) deceptive. The connection between them is so close that we are to regard them as seamlessly welded together into a four-dimensional spacetime. There can be no time without space, nor space without time.

The significance of this in relation to what we have just said about the big bang should be clear: If the instant of the big bang marked the origins of space, it also must have marked the origins—the beginning—of time. There was no time before the big bang. One cannot talk about what happened before the big bang because there was no "before."

At a stroke, this renders untenable the commonly held idea that in the beginning there was God; initially he existed alone, and then at some point in time he set to and made the world. Indeed, according to an idea put forward by Stephen Hawking, there might not even have been a first instant of time.[3] It could be that as one imagines going back in time toward the beginning of the world, time itself ceases to be time; it gradually "melts away," becoming more and more like the other three (spatial) dimensions. It was thoughts of this nature that prompted Hawking's remark quoted earlier: "What place then for a creator?"

But *is* this development in cosmological thought as damaging to the notion of a Creator-God as it appears to be? Surprisingly, none of these considerations regarding time is new to theologians. At least 1,500 years before cosmologists stumbled on these ideas, St. Augustine had already satisfied himself that there could not have been time "before" the world had been created.[4] He, of course, knew nothing of the big bang and the type of thinking that has led modern cosmologists to this conclusion. Instead, he argued that the way we know that there is such a thing as time is by observing change—the way objects change their position relative to each other *in time*. He said that if there were no motion—if everything remained stationary and had always done so—the word "time" would refer to nothing and would therefore be meaningless. If indeed there were no objects at all (let alone moving ones) because they had not yet been created, then that would be an even stronger reason why there could be no time. Time, according to Augustine, was a property of the world and as such must have been created along with everything else. And yet none of this affected his faith in a Creator-God. Why?

The important point is best made by drawing a clear distinction between two terms that are often used interchangeably: creation and origins. If one is interested in how everything started, then one is dealing with *origins*. On the other hand, if one is asking, Why is there something rather than nothing? Why is there something *now* and at all other times?, then that is a matter of *creation*. God as Creator gives existence to the world moment by moment. That is why traditionally theologians inextricably link the notion of God the Creator with the equally important and relevant one of God the Sustainer.

The big bang origin of the universe and its subsequent development through time rightly falls within the domain of the scientist. The question of creation, on the other hand, is seeking the ground of all being—the source of existence of all space and time, and as such, falls within the province of the theologian. In this latter context, what happened at the first instant (or indeed whether there was a "first instant") is quite immaterial.

God of the Universe

Such, then, are some of the considerations arising out of modern cosmology. But long before these ideas about the origins of the universe were formulated, astronomy had been forcing a reassessment of our ideas of God and of our own place in the cosmos.

In the light of the work of Copernicus and Galileo, it came to be recognized that, contrary to previous assumptions, Earth did not occupy center stage; it was but one planet among several orbiting the Sun. This process of reevaluating the importance of our earthly home has continued apace ever since. Our present understanding of the immensity of the universe almost defies description. One million Earths could fit within the volume of the Sun. Each star is a sun. There are 100,000 million stars in the Milky Way galaxy. There are 100,000 million galaxies. They are spread out over such volumes of space that it has taken light 12,000 million years to travel to us from the farthest reaches of space—and that despite a speed of 186,000 miles per second.

And God created all this. We have indeed come a long way from the God who was originally thought to be confined to living on a mountain in the Sinai desert! Contemplation of such vastness, of the awesome power of supernova explosions that mark the cataclysmic death of giant stars with an outburst of light that can outshine the light of a whole galaxy, and above all the sheer violence of the big bang, puts a whole new emphasis on what we mean by saying that God is all-powerful. The psalmist who saw the heavens as declaring the glory of God could have but glimpsed the true splendor that lay beyond the few thousand pin pricks of twinkling light visible to the naked eye that was his idea of the heavens.

And what of life elsewhere in the universe? Already there is firm evidence pointing to seven cases of planets going round nearby stars.

With the new observational techniques now at our disposal, we are doubtless on the threshold of discovering many, many more worlds. From what we know of the formation of stars and planets, we would expect there to be vast numbers of planetary systems similar to our own solar system. A substantial fraction of these are likely to include planets so placed in relation to their sun that they would be temperate and thus potentially conducive to the development of life.

Whether life actually *has* developed there or not is a matter of conjecture. Until searches for signals from outer space succeed in demonstrating an alien culture trying to make contact with us with intelligent messages, there can be no proof one way or the other. But to assume meanwhile that we are the only intelligent life form in the universe would appear somewhat egocentric. Having learned a lesson in humility from Galileo, perhaps it is more realistic to accept that the universe probably teems with intelligent life forms.

That being so, one can imagine that God is as much interested in those aliens as he is in us—indeed possibly more so, if they have advanced spiritually more than we have so far on this planet. All of which, of course, is a far cry from the tribal god interested only in the Jews—or even the one who was simply the God of all the people on Earth. There is no doubt that the findings of astronomy have done much to draw our attention to the sheer scale on which God works.

The God of Evolution

Earth formed some 4,600 million years ago, and it has taken that long for evolution by natural selection to transform inanimate chemicals into human beings. In fact, it has taken a total of 12,000 million years since the big bang to produce us. A long, long time. From the vantage point of this modern understanding, we have a far better appreciation of the patience and far-sightedness of God than was possible in earlier times. From God's perspective how petty our own restive, anxious demands for immediate results and instant gratification must seem.

Second, we learn from evolution that God is willing to incorporate into the achievement of his purpose an element of chance. The mutations on which natural selection works occur randomly. There is no conscious, detailed "design" built into all the minutiae. There is a hands-off openness about the process. God appears to be allowing his world to be itself. And yet he knows that the whole system has been set up in such a way that his broad aims eventually will be achieved. Intelligent life of one form or another will in due course emerge.

On other planets, the random events giving rise to life will be different, so humans as such will not evolve. But *some* form of intelligent life eventually will appear—a form of life that will at some stage begin to ask the ultimate questions concerning the purpose of life and whether there exists anything beyond the visible and tangible. At that stage, a relationship with God is established, which is why the world is here in the first place. In these ways, our conception of God has to broaden.

One particularly difficult problem inherent in evolutionary theory has to be mentioned. Evolution by natural selection has been characterized by the phrase "survival of the fittest." It is a crude description but has some truth to it. It is all about survival, as is plain to see from the constant engagement between predator and prey. Although it is impossible to be sure what any other animal actually *feels*, it seems only reasonable to conclude that evolution involves suffering on a massive scale.

Now, of course, there is nothing new about suffering. The problems of evil and suffering have been with us since time immemorial. Evil can be accounted for in terms of our own disobedience against the wholly good God, and some suffering arises directly out of evil acts or from wanton unwillingness to help those in need. But not all suffering arises in this way; even in the absence of human wickedness, there is suffering through natural causes such as earthquakes, flood, and failure of crops.

A partial explanation might lie in the need for rigid laws of nature to hold sway so that we can exercise free will in an environment in which we know what the outcome of our actions will be. Inevitably, one will sometimes fall foul of those laws working out their inexorable course. One also has to accept that in a hypothetical world where there was no suffering, it would be difficult, if not impossible, to demonstrate one's love for another (in the way one attends to his or her needs and is prepared to make sacrifices on that person's behalf). But having said that, the sheer *degree* of suffering in the world always has appeared to some to be excessive. Now, on top of that, we have to come to terms with the fact that the very process by which intelligent beings evolve incorporates by its very nature an unavoidable degree of intense suffering. Why did God choose evolution by natural selection? Was there no other way? Clearly, we still have much to learn about the mind of God in this matter.

God of Time

In discussing the big bang, we saw that space and time are much more like each other than we might at first have suspected. The view we get from Einstein is of a four-dimensional spacetime rather than a three-dimensional space that evolves in time. It is a rather strange picture of reality. Nothing changes in four-dimensional spacetime. In order for there to be change, it must take place *in* time. But time is part of the four-dimensional reality we are talking about; there is no other. For this reason, many physicists talk of a "block universe," one that contains all of space and time on an equal footing. So, for example, just as all of space exists at each point in time, so all of time exists at each point in space. Physical time is essentially static.

This is not to say that space and time are indistinguishable. They clearly are very different, as we noted in the ways in which we measure the two: a ruler in the one case, a clock in the other. Not only that, whereas one can travel no distance in a finite time (just stay still), one cannot travel a finite distance in no time. To do the latter, it would have to be feasible to travel at infinite speed.

This possibility is ruled out by another consequence of relativity theory; namely, nothing can travel faster than the speed of light (300,000 kilometers per second). This constraint means that if one represents the motion of an object through space and time by tracing out its path in four-dimensional spacetime, it can lie as close as it likes to the time axis (by remaining stationary, it would lie exactly along this axis). But there is a limit to how closely it can align itself with any of the spatial axes (the closest alignment being that of a light beam). So, a study of these paths in spacetime would single out the time axis as different from the other three.

In addition, we have the second law of thermodynamics: Disorder increases as time increases. If we have, for example, a photograph of an intact cup and another showing the same cup smashed, we immediately know that the latter relates to a set of circumstances found at a *later* time; in other words, nearer to the positive end of the time axis. The spatial axes do not exhibit any such asymmetry.

So far we have talked purely in physical terms: the motions of objects and light, measurements recorded by physical rulers and clocks, orderly and disorderly states. When we add to this our experience of space and time *as conscious human beings*, then further differences arise. In particular, as regards time, we become aware of a distinction among past, present, and future. We seem to inhabit the instant known as "now"; the past no longer existing; the future yet to exist. Moreover, we are aware of the "flow" of time. We move toward the future; we do not move toward the past (outside the realms of science fiction, that is).

Whereas we noted that according to the second law of thermodynamics there was more disorder toward one end of the time axis than toward the other, this conscious experience of time takes us in the direction that leads to more and more disorder. This inexorable movement is a feature only of time; there is nothing equivalent to compel us to travel in only one spatial direction.

We are thus confronted with two entirely different understandings of time. On the one hand, we have conscious experience presenting us with a flowing time in which the constantly moving special instant called "now" separates the totally different domains of past and future. On the other hand, physics presents us with a static time in which no instant is singled out as in any way special, all instants of time being on an equal footing, just like all points in space. How are we to reconcile the two? How are we to see God in relation to time?

There is no easy answer. The view of time that is most readily grasped is, of course, that of conscious experience. The static idea of time is so alien and counterintuitive that even some professional scientists are inclined to dismiss it as being somehow wrong. In contrast, there are others who would claim that the scientific picture has to be correct, and it is our conscious experience of time that is illusory. But it seems to me that rejecting, or downplaying, one of these approaches to the understanding of time in favor of the other is not the right course of action.

Instead, we have to come to terms with a mystery that defies our normal categories of thought and our usual ways of organizing information. The dilemma over the two kinds of time points us toward a different *type* of understanding, one in which we have simultaneously to hold in the mind seemingly contradictory, paradoxical conceptions, each of which embodies some of the truth but not the whole. (We saw something of this when we held in the mind simultaneously the ideas of God being both our father and our mother.)

To get the complete picture of reality, one needs both conceptions. One has to learn to accept that this conjunction of seemingly paradoxical ideas *is* the explanation. It might not be the type of explanation one had been expecting, but no matter. One must allow the nature of reality to dictate not just the answer to our question but also the very form of the question we ask and the form of its answer. Constraining the outcome to fit in with our conventional notions as to what constitutes a satisfactory "explanation" can only distort the truth and leave us with an impoverished understanding. We have more to say on this in the next section when we tackle quantum theory.

But, for now, we note that our findings about time seem to point us in two directions at once. This in turn affects the way we ought to see God in relation to time:

God is to be found *in* time. He participated in human life through Jesus Christ. He is the God to whom we pray. For prayer to be effective it must bring about change. We must, therefore, live our lives on the assumption that God does react to us; not only do we change with time but to some extent, and in some sense, he, too, changes—in answering our prayers.

But he is not just the God of mental life with its experience of flowing time and change. He is also the God of the physical world—

the four-dimensional integration of all space and all time. As such, all of time is present to him; he sees it all; he knows it all. This is the aspect of God's relationship to time that goes against the grain. It is difficult not to incorporate into our picture of God an extension of our own human limitations on knowing the future. But if we are to reach out to a more sophisticated understanding of God's nature, this is a temptation that has to be resisted. We must allow whatever are the currently accepted best interpretations of science to be the arbiter and guide. And the fact that those interpretations appear to include a time such that, in some sense or other, all of it exists on an equal footing, surely makes it easier to accept that God has knowledge of the future.

Not, of course, that there is anything new about the idea of God knowing the future. It is all there in the teaching of St Paul, for instance. What modern science does is to lend an added measure of credibility to this ancient insight.

The acceptance of God's foreknowledge as something that arises from God's ability to encompass the whole of physical time from a vantage point lying beyond such time itself, feeds back into our own personal relationship with God occurring within the type of time that is relevant to conscious experience. The God to whom we pray in that context is the God who *knows* what the outcome will be and who *ensures* that all will be well. In the same way, the God who built into the evolutionary process a measure of random chance goes beyond a God who has simply stacked the odds so that it is overwhelmingly likely intelligent life of some form will appear somewhere; he knows *exactly* what form that intelligent life will take.

The idea of a God holding the somewhat paradoxical engagement with time that is both within the changing, open-ended time of conscious experience and also beyond the unchanging completeness of physical time is not easy to grasp. Indeed, let us be frank: It *cannot* be done. It cannot be grasped in the same sense as one might be able logically to prove a geometrical theorum. It is a truth, an understanding, of God that is to be *accepted* rather than mastered. It is a paradox that *points* to the truth, rather than *encompasses* it. This paradoxical approach affords a way of further deepening our appreciation of the nature of God beyond that which can be achieved by the straightforward use of any single metaphor drawn from everyday human life.

The Trinitarian God

The use of paradox in explanation is not to be taken as a licence for sloppy thinking. One cannot just throw together any set of opposing ideas. A good example of the disciplined use of paradox (i.e., seeming contradiction) is to be found within the realm of quantum physics in the so-called "wave/particle paradox."

Let us, for instance, take the behavior of light. When it interacts with matter, it behaves like a tiny particle. The collision takes place instantaneously, at a particular pointlike location, with the exchange of energy and momentum. For this purpose, we regard light as a packet of energy and momentum, called a *photon*. A beam of light will consist of many such photons. Where the next photon interaction will occur, we cannot tell. All we can do is compute probabilities of various alternatives: the quantum uncertainty of which we earlier spoke. To evaluate the probabilities we have to treat the light beam not as a stream of particles but as a wave. Accordingly, the passage of the beam through slits in screens, troughs of water, glass prisms, and lenses, etc. shows all the characteristics one associates with waves: diffraction, reflection, refraction, and polarization. Thus, we are able to work out the intensity pattern of light falling on a distant screen, and this in turn tells us the relative probabilities for finding the next photon arriving in any given region of the screen.

Now, this is very strange. The photon is envisaged as having no size; it is pointlike in its interaction. But a wave by its very nature *cannot* be pointlike; it has a wavelength and that has to be spread out over a certain distance in space. So we are faced with the problem of how something can be localized to a point and yet *not localized*.

This is a problem that affects not just light but everything else: electrons, protons, nuclei, atoms, and even molecules. Although they interact with other forms of matter as though they were localized particles, in computing *where* they were likely to interact, they have to be regarded as waves—waves possessing a wavelength governed by their momentum. It is merely the fact that large "molecules" like human bodies, footballs, and cars have large values of momentum, and hence tiny wavelengths, that disguises their wavelike characteristics. The truth, however, is that every time one walks through the slit that is a doorway, for example, one gets very slightly diffracted to one side or the other.

Thus, we are presented with the wave/particle paradox: how something can be localized and yet *not* localized. But it is not a free-for-all. One knows exactly when to invoke the wave behavior (in computing probabilities) and when to invoke the particle behavior (at the instant of interaction). To this extent, everything is cut-and-dried. With this manner of working, one is able to understand the properties of atoms and design and build lasers, transistors, supercooled magnets, etc. Quantum physics has become a thoroughly applied science, a branch of technology.

The trouble comes if one wishes to go beyond the practical day-to-day exploitation of quantum physics to ask questions such as What actually *is* light? or What actually *is* an electron? Here, for instance, one is looking for a straightforward, common sense answer to the question as to whether light is *really* a pointlike photon—a particle that just happens to look a bit wavy at times (whatever that means)—or whether it is *really* a spreadout wave—a wave that just seems a bit gritty from time to time (whatever *that* might mean).

The solution offered by Niels Bohr was to reject this "What is" type of question as meaningless.[5] We are not to talk of what something *is*, rather we are to restrict our discussion to how things *behave*. As long as we talk only of behavior, i.e., of interactions, there is no problem. If we are asking *how* light or an electron interacts, we use the language of particles; on the other hand, if we ask *where* the light interacts, we use the language of waves. What we are *not* permitted to ask is what the light is outside the context of an interaction (i.e., outside the context of its observed behavior).

Bohr argued that all the words we use belong to a language concerned solely with interactions. It is, therefore, a *misuse of language* to apply it to what might or might not exist, or what might be happening or not happening, *in between* the instants of those interactions. Thus, we can talk of properties such as "energy" and "momentum" and "position" as characterizing certain features of an interaction, but that is not to say that it makes sense to think of an electron as having, say, a "position" or an "energy" in between its interactions.

According to Bohr, the most complete picture of an electron, for example, is not really as a particle (downplaying the wave characteristic) or alternatively really as a wave (this time downplaying the significance of the particle nature). Instead, one must hold simultaneously in mind both the particle and the wave aspects, according each equal status. It is through a combination of identifying which are meaningful questions and which are not, together with an acceptance as to what constitutes a satisfactory "explanation," that one gains a resolution of the wave/particle paradox.

Although this interpretation of quantum physics has come to be known as the orthodox one, not everyone, by any means, goes along with it. There are those who believe this is no kind of explanation at all and that Bohr and his followers have given up too easily on the problem through, in effect, denying that there remains a problem. Those dissatisfied with the so-called Copenhagen interpretation (originated by Bohr) have included many eminent scientists, including Einstein.[6] But 60 years from the time this debate was initiated, no-one seems any closer to an acceptable solution as to what might be going on in between interactions, which is exactly what Bohr would have expected.

Whether this is, or is not, the right interpretation of quantum findings, one thing is for sure: The kind of thinking that Bohr employed appears to have its uses in theology. One has only to think of the Christian doctrine of the Trinity to be faced with paradox: the three-in-one and one-in-three. Or again take the Christian understanding of Jesus as both fully human and fully divine. He was not to be regarded as *really* God and only appearing to be human; neither was he to be thought of as *really* human and only appearing to act in a godly manner. He was *fully both*, with neither emphasized at the expense of the other. And this was despite having to marry the apparently contradictory requirements of, on the one hand, God being all-powerful, all-knowing, and unlimited as regards space and time and, on the other, a human being weak, limited in knowledge, and restricted to a finite location and a narrow span of time.

Despite the contradictions involved, the early Church fathers felt compelled by their interpretation of the life, death, and Resurrection of Jesus, his teachings, and all the events that had presaged his coming, to embrace these paradoxical statements as to the nature of God and the nature of Jesus. It is extraordinary that the same kind of thinking that 1,500 years ago assisted in the resolution of these questions concerning the Creator should have resurfaced in our own time as the way forward in understanding the Creation.

The Way Forward

Now, of course, it can be argued that my past two sections have not advanced theology at all in that God's foreknowledge, the thinking behind the trinitarian formulation, and the idea of Jesus being both fully human and fully divine—none of these is new. That's true. But progress sometimes can be made by regaining lost ground. With the growth of rational thinking, a premium has been placed on straightforward, logical, and (supposedly) scientific explanation. Thus, it is common to come across professional theologians who do not subscribe to the idea that God has foreknowledge. For them, it is an impossibility—and even an all-powerful God cannot be expected to do things that are logically impossible.

What we find in the light of relativity theory is that one interpretation of the nature of spacetime, widely held by many scientists, is that the existence of the future is far from being an impossibility—it is a reality. The fact that we human beings do not have the facility to gain access to knowledge of that future is completely beside the point; the future *is* there. That being so, the current trend among some theologians to discount God's ability to encompass that future smacks of fashioning one's conception of God to fit in with human limitations. Perhaps this is where modern physics can play a useful role in recalling us to a former, and more adequate, understanding of God.

Likewise, there is a trend to discount trinitarian thinking and the doctrines concerning the person of Jesus as having been too heavily influenced by Greek thought. Bowing to the seeming demands of our modern scientific culture to abandon the idea of miracles (at least in the sense of their being departures from the normal operation of the laws of nature), there is a move to reject, for example, the virgin birth. Whether this is justified or not (and on balance I would incline to the view that the account of the conception of Jesus probably ought not be taken as literally true), does this mean we should abandon the claim of the divine nature of Jesus? It certainly does not necessarily follow. The facility with which the modern quantum physicist simultaneously sustains a balance between the seemingly contradictory wave and particle properties of the electron should serve as an example to theologians that there might still be much to be gained from continuing to do likewise over the human and Divine natures of Jesus.

And just as this paradoxical type of explanation stemming from science can serve as a useful antidote to overzealous theological attempts to "rationalize" traditional doctrines, so it might be pointing the way forward for further explorations into the nature of God and the way he relates to us and to the world. For example, what are we to make of the notion of freewill once we accept that our choice is somehow frozen into the "block universe"; this appears to involve a contradiction of some kind. Second, in the absence of miraculous interventions into the workings of nature (or at least if interventions occur with considerably less frequency than was once thought) how *does* God have an effect on the world? Third, there is the old problem of why there should be injustice, evil, and suffering in a world made by a God of justice, goodness, and love—a problem which, as we noted earlier, is seemingly exacerbated by the recognition of the intense suffering incorporated into the very process by which we were formed: evolution by natural selection.

It appears that progress with any of these questions must somehow incorporate within them paradoxical modes of thought—the new kind of thinking that has now surfaced in science and has always had a place in Christian theology. In seeking yet deeper insights into God, we must be alive to the possibility of radically different *kinds* of explanation—otherwise we run the risk of achieving that improved understanding, but not recognizing it when we've got it.

Notes

1. Sigmund Freud, *The Future of an Illusion* (London: Hogarth, 1927).
2. Stephen Hawking, *A Brief History of Time* (New York: Bantam, 1988) p. 141.
3. Ibid., p 139.
4. St. Augustine, *The Literal Interpretation of Genesis*, Book 5, chap. 5, 12.
5. M. Jammer, *The Philosophy of Quantum Mechanics* (New York: Wiley, 1974).
6. Ibid.

How Large Is Faith?

HERBERT BENSON AND MARG STARK

Starting out in the profession of medicine in the United States more than thirty years ago, few questions would have struck me (Herbert Benson) as more irrelevant to science than the one posed by the title of this book, How Large Is God? After all, I was indoctrinated, as all Western physicians are in medical school and in my subsequent training, to believe that religion was the polar opposite of science. And it seemed to me, early in my career, that life's mysteries were, one by one, being explained by modern technology and research. So I became impressed with the enormity of science, not with the enormity of God. (When referring to God, the authors mean to describe all the deities that people believe in and worship.)[1]

In this posture, I was as wedded to conventional wisdom as my colleagues, all of us descendants of René Descartes, who, among others, forged the fissure between mind and body and between spirit and biology that prevails in Western scientific thought today. But as I began to practice medicine, I noticed that when it came to identifying sources of healing, my colleagues and I were often out of sync with patients. For so many of my patients, God and faith in God were not only relevant but real influences in their lives and in their health. Even so, I would not have ventured into a realm as unscientific as religion had it not been for mind/body research that led me to wonder if faith in God could be healing in and of itself.

Thus began my investigation into these largely unexplored resources, the resources on which patients relied every day of the year, sometimes consciously but often unconsciously, without the help of physicians or pharmaceuticals. My investigation meandered, as all scientific pursuits do, from experiment to research results, from new hypothesis to another experiment. But eventually, by way of science, I was compelled to look seriously at faith and at the magnitude of God as seen through the eyes of my patients.

The Calm Within

I came full circle and began to consider the influence faith in God could wield in the body, only after a prolonged scientific quest, the milestones of which I will now review. I was a young cardiologist when I received one of my first clues that something in the traditional approach to patient care was amiss. In those early years, I was troubled that by using a single test—the measurement of blood pressure—a physician would diagnose hypertension and prescribe a lifelong regimen of medicines with considerable side effects. However, before coming to see the doctor, the patient may have felt perfectly well and may not have experienced any other symptoms of serious illness. Hypertension, or sustained high blood pressure, is often symptomless but can have many detrimental, long-term effects, such as blocked arteries and enlarged and strained hearts, and can cause a variety of heart disorders and strokes.

Moreover, the readiness with which we physicians turned to prescription drugs bothered me. Not only were the side effects bothersome to patients but the medication often caused hypotension—blood pressure that is too low. This pattern suggested to me that patients' blood pressures may have been artificially elevated by the stress of visiting their physician. I thought that perhaps hypertension could be brought on by stressful events, a correlative that had not been fully recognized at that time.

To determine whether stress affected blood pressure, I began experimenting with monkeys in a Harvard Medical School laboratory and found that the monkeys, responding to certain incentives and disincentives, could actually learn to control their blood pressures. The monkeys required no medicines to increase or reduce the force of the blood that pumped through their arteries.

According to this study, stress could contribute to high blood pressure. And perhaps more important, this study made it seem possible that humans also could be trained to control and lower their blood pressure.

Transcendental Meditation

While the aforementioned experiments were underway, I was approached by some practitioners of Transcendental Meditation who believed that they were capable of lowering their blood pressures when they meditated. TM, as it was known, became popular in the late 1960s and early 1970s, under the leadership of Maharishi Mahesh Yogi. But I turned the TM practitioners away. My findings that stress might contribute to high blood pressure already had met with resistance at Harvard and elsewhere within the medical community, and I was fearful of jeopardizing my primate research by associating with what others deemed a "fringe organization."

But the TM practitioners were persistent, and eventually I brought some of them into the laboratory to measure physiologic changes that occurred while they engaged in meditation. Amazingly, the results proved them right. During their meditative exercises, they experienced low blood pressures, as well as significant reductions in heart rate, breathing rate, and speed of metabolism.

Ultimately, I called this physiologic calming effect "the relaxation response." Researchers before me already had discovered that this same kind of physical relaxation could be induced in animals. Nevertheless, the findings that TM could be used, as a willful exercise, to produce physiologic changes in humans sparked controversy in the medical community. While some of my colleagues were supportive, others urged me to abandon this line of research. Still others advised my superiors not to allow me to accept grants dedicated to this purpose, fearing the harm my association with a "cultish" element would do to Harvard's reputation.

In the end, it was Harvard's prestige and the open-mindedness of Dr. Robert H. Ebert, former dean of faculty of medicine at Harvard

Medical School, who enabled my research to continue. The dean decided, "If Harvard can't take a chance, who can?" Thus, I was allowed to accept grants and to continue my studies of the relaxation response.

The Placebo Effect

One of the arguments used to dismiss the relaxation response was that it was "nothing but the placebo effect." The placebo effect describes a phenomenon in which a patient's beliefs and expectations about a pill or therapy contribute to the effectiveness of the treatment. Less well known is that a caregiver's beliefs and expectations, and the beliefs and expectations generated by the relationship between caregiver and patient, also contribute to the effectiveness of a pill or therapy.

Today, medicine primarily employs the placebo effect in new drug trials in which a control group is given inactive pills as a source of comparison with the medication being tested. But the theory behind the placebo effect that physicians have long recognized is that having faith in something makes a difference in medical outcomes. Traditionally, we have relied on the findings of Dr. Henry K. Beecher of Massachusetts General Hospital, reported in 1955, that the success rate of treating various types of illness with the placebo effect is roughly 30 percent.[2]

But when I looked closer, in an attempt to discredit the placebo's role in the bodily calm I had observed, I was instead impressed by how much more powerful and widely underrated the placebo effect was in medicine. My colleague Dr. David P. McCallie Jr. and I looked at therapies used in the past to treat angina pectoris, pain that people with reduced blood flow to their heart muscles experience in their chest and arms. All the treatments have since been disproved and abandoned, and many of them, from injections of snake venom to the removal of the thyroid, sound downright absurd to modern ears.

And yet, Dr. McCallie and I found that when the treatments were in vogue, and were *believed in*, these physiologically indefensible and inherently worthless therapies for angina were effective 70 to 90 percent of the time. As these same treatments gradually fell out of favor and as physicians began to question their use, the effectiveness of the procedures indeed dropped to 30 to 40 percent. When patients and their doctors *had faith in* the therapies, the success rate in alleviating angina was two to three times the rate that Dr. Beecher attributed to the placebo effect.

As medical research amassed evidence that the mind and the body were interrelated, and as the mechanisms responsible for the placebo effect became clearer to me, I renamed the effect "remembered wellness." Not only did remembered wellness better describe the brain's influence over physiology but I hoped that a new name would help medicine think about the phenomenon differently. After all, the placebo effect had become a pejorative, like "the dummy pill" and "it's all in your head," terms that doctors used to dismiss a patient's beliefs or emotions.

Dr. McCallie and I concluded that medicine seriously devalued remembered wellness as a therapeutic tool. But it took more than twenty years for the message of faith's *real* influence to find an audience in the medical community. In the meantime, I returned to elucidating the relaxation response, the study of which eventually led me back to the powers of remembered wellness.

The Relaxation Response

The next step in establishing the existence of the relaxation response was to explore whether TM alone could produce these physiologic changes, or whether other techniques could be used to elicit the relaxation response. TM involved two basic steps that were neither complicated nor mysterious. First, a person had to focus his or her mind on a word, phrase, or sound and to passively disregard interfering thoughts and return to the focus. After three to five minutes of this repetitive mental focus, the relaxation response and its corresponding reductions in heart rate, breathing rate, rate of metabolism, and blood pressure occurred. Soon thereafter, I found that anyone who employed these two steps could elicit the physical changes of the relaxation response.

My colleagues and I, working first at Beth Israel Hospital in Boston and later at the Deaconess Hospital where I helped establish Harvard's Mind/Body Medical Institute, discovered that not only was the relaxation response good for people in the short term but it also had magnificent long-term preventive and restorative properties for people who elicited it regularly. In addition to hypertension, we found that the relaxation response either cured or lessened the effects of chronic pain, insomnia, infertility, premenstrual syndrome, anxiety and mild depression, headaches, and low self-esteem. The relaxation response also successfully reduced the nausea associated with chemotherapy, the pain and anxiety associated with surgery and X-ray procedures, and the frequency of cardiac arrhythmias in heart patients. And demonstrating its economic value, the elicitation of the relaxation response along with other stress-management techniques reduced the number of visits patients made to doctors and improved worker attendance on the job.

The relaxation response seemed to cure or help any medical condition or illness to the extent that condition or illness was caused or exacerbated by stress. Because this physiologic state of calm was accessible to everyone, I became convinced that the relaxation response was the opposite of, and perhaps the antidote for, the stress-induced fight-or-flight response.

Identified by Dr. Walter B. Cannon, the fight-or-flight response is the body's mobilization for either battle or escape, crucial to the survival of our distant ancestors and remaining a part of our genetic equipment today.[3] The physiologic changes that occur in the fight-or-flight response, namely heightened blood pressure, heart rate, breathing rate, and an increased metabolic rate, are precisely the opposite of those that occur during the relaxation response.

Even though we inherited the fight-or-flight response because it was good for our ancestors, it is not necessarily as good for us in modern life. We do not usually expend the physical energy that may be called forth in us when we are threatened. Unlike cave men and women, our everyday threats are not usually physical. And the long-term effects of the repeated triggering of the stress-induced fight-or-flight response are injurious to our bodies.

On the other hand, the relaxation response could restore a body's equilibrium, offsetting the long-term damage that repeated stress and the fight-or-flight episodes caused. Although our genetic heritage, indeed our wiring, often makes stress a negative influence in our bodies in modern life, we also are "wired" to experience the physical balm that my colleagues and I had proven was healthy.

The Historical Perspective

To see whether or not the relaxation response had served this purpose for ancient humans, I set about looking for historical evidence of its presence. Afer two years of studying the literature of the world, I was thrilled to find that in every time, and in every culture in recorded history, there were examples in which humans described calming techniques. These methods incorporated the two essential steps—a repetitive mental focus and passively ignoring any interrupting thoughts to return to one's focus—and were indeed practiced in every religion, and sometimes outside of religion, in every nationality known to us.

I was pondering all this, some thirty years ago, as I mulled over the wealth of documentation I had accumulated about the relaxation response's physical value. I thought about how people in different cultures and times described the same physical relaxation that I had observed in my patients' experiences. And I considered what it meant that we were wired to experience this physical balm, accessible by mental focusing.

Suddenly, I came to a startling conclusion: This is prayer! The quiet ritual that people had observed in many different ways in many different lands, and that religious leaders always said was good for people, seemed to call forth a particular set of bodily changes, opposite of those called upon in a crisis. People seemed to internalize and embody the repercussions of a repetitive form of prayer in ways that nourished them and promoted their physical health, not just their emotional and spiritual well-being.

The Study of Prayer

As true as this realization seemed to me, it was a conclusion I came to dread. Because, aside from being controversial, the study of prayer or any religious subject was considered unscientific. Nothing in my training as a physician prepared me to measure the effects God, or gods, had on my patients. As undeniable as it was that people throughout history had practiced and physically reacted to repetitive prayer in ways that I could prove were beneficial to the body, there was little or no scientific precedent for studying faith in God or religious traditions as healing methods.

Equally, I feared the reaction of the religious community. Would it appear to believers that I was trying to quantify God and to reduce the effects of religious faith to what I could see under a microscope? With this in mind, I went to see the dean of Harvard's Divinity School, Dr. Krister Stendhal. Dean Stendhal was a tall man, especially to me, as I am five foot nine, even with my best posture. But his presence was bigger still, or so it seemed to me as I reviewed the points I had rehearsed.

I told him about the relaxation response and about the common experience of calm brought on by repetitive prayer as documented throughout history. I told him that in the process of distinguishing the relaxation response from the placebo effect, I had been impressed by the role belief and faith played in physical processes, a role misunderstood and underappreciated by modern medicine. And I told him I wanted to study the role of prayer, and specifically the mental focusing aspects of prayer, as well as the influence of faith. But I was worried about the effect my scientific study would have on believers and on the steadfastness of their faith.

Dean Stendhal listened closely to everything I said before he rose to deliver a reply. Looming above me, as I sank deeper into my chair's soft cushions, Dean Stendhal answered my queries directly, "Young man, don't worry about us. Religion and prayer were here before you and they will be there after you. You do your thing and we will do ours."

Spirituality Linked to Health Benefits

And so it was that I began the delicate study of repetitive prayer, of the faith in God that usually accompanied it, and of the extent to which both helped people in measurable, medical ways. The effects of the relaxation response already had been established. Non-religious people and those who approached mental focusing as a health-enhancing exercise not as a spiritual one experienced physiologic changes and benefits just as religious believers did.

But as much as traditional medicine urged me to exclude faith from consideration, 80 percent of my patients chose to say a prayer to elicit the calm of the relaxation response. Because my patients were naturally drawn to prayer, I found myself in the awkward position of being a physician who taught people to pray, despite the fact that I encouraged people to use any repetitive word, phrase, image, or even a repetitive activity, such as knitting or jogging, to focus their minds and bodies. My only stipulation was that it be something that patients liked and felt comfortable with, so that they would be more apt to adhere to the routine of eliciting the relaxation response.

Not only did most patients choose scriptures or prayers but 25 percent of those who elicited the relaxation response experienced an increase in spirituality, whether or not they intended to. Dr. Jared D. Kass, a professor at Lesley College Graduate School of Arts and Sciences in Cambridge, Massachusetts, other colleagues, and I developed a questionnaire to pin down what people meant by "spiritual" when they talked about the effects of mental focusing. Again, whether or not the practitioners considered themselves religious, those surveyed said two things about this sense of spirituality: They felt the presence of an energy, a force, a power—God—that was beyond themselves, and this presence felt close to them.

In addition, the 25 percent who felt more spiritual as the result of eliciting the relaxation response experienced fewer medical symptoms than those who did not experience spirituality. This amorphous sense that we had nailed down—the presence of something beyond them that nevertheless felt close to them—seemed to lend their bodies additional healing effects, over and above those of the relaxation response.

A Dearth of Research

This should not have come as such a surprise. After all, something as time-honored in people's lives as faith surely must have been dissected and quantified by medicine before. According to a 1990 Gallup poll, 95 percent of Americans said they believed in God, and 76 percent said they prayed on a regular basis. How was it possible that Western medicine had missed something as integral to life, and as seemingly therapeutic, as faith?

I researched this question while reviewing the medical literature for confirmation of what I had found—that faith could heal. And I was disappointed to learn that Dr. Robert D. Orr and Reverend George Issac of the University of Chicago, in their review of 1,066 medical journal articles, found only 12 that assessed religious factors, a mere 1.1 percent of the total.[4] Indeed, Drs. Jeffrey S. Levin and Preston L. Schiller of Eastern Virginia Medical School found only 200 studies that included religious findings among the hundreds of thousands of English medical journal articles published over the past 200 years.[5]

Nevertheless, the findings that were available demonstrated the value of religious belief and practices for a person's health. And in the 1990s, scientists launched additional studies of religious factors. For example, Dr. Thomas E. Oxman and colleagues at Dartmouth Medical School found in 1995 that people over the age of fifty-five who had open-heart surgery were three times more likely to survive if they reported receiving solace and comfort from religious beliefs.

It did not matter which religion. Many aspects of religious involvement, including voluntarism and fellowship, proved to be very good for people's health. Overall, belief in God lowered death rates and increased health, according to Dr. Levin, who reviewed hundreds of epidemiologic studies to make these conclusions.

Dr. McCallie and I had hinted at this years before, when we learned that remembered wellness, and the catalyst of belief, was two and three times as powerful than medicine typically acknowledged. But my subsequent studies of the relaxation response, of the natural tendency of people to add their religious beliefs to the practice of mental focusing, and of spirituality's additional health benefits made the cause of faith in medicine undeniable to me.

I became far more interested in the combination of remembered wellness and the relaxation response than in isolating them. I termed their combined use "the faith factor." And I surmised that the reason that the faith factor was particularly potent was because we are mortals and recognize that our bodies eventually will succumb to illness and death. So, in order to fuel remembered wellness, we often need to believe in something bigger and stronger than we are. Believing in a higher power helps us overcome our doubts and fears and to put aside anxieties that would otherwise inspire the fight-or-flight response. That way, we can mentally envision health for ourselves and send important signals to our bodies.

The Modern Appreciation of the Brain

The modern understanding of the brain supported this hypotheses. Because the latest brain research revealed that it was far more complex than medicine had appreciated, and because it appeared that mind and body were inseparable, interconnected entities, all our attempts to separate them artificially began to seem foolish.

For example, scientists once thought that the brain received messages primarily from the body and the external environment—from the bottom up. Let's say one smells some freshly baked muffins. Nerve cells in the nose send a message through the body to the mindful regions of the brain where the "muffin-smell alert" will be processed. The brain's reaction to this news will then be fired off along nerve cells so that one might begin to salivate. This is a bottom-up event.

Top-Down Events

However, recent studies have indicated that one could sit right where one is sitting and imagine the smell of a muffin, even though there isn't one baking nearby, and might again salivate. And in the process, the same nerve cells would be activated as would occur if one actually did smell a muffin, fresh from the oven. This is called a top-down event because the message started not from the body, not from the external environment but from the top, in the imagination and in the mind.

The repercussions of top-down events for our understanding of the brain and body are enormous. And both imaginings and memories can produce top-down events, as I experienced in a very powerful way years ago, when I was an intern at a hospital in Seattle. I was working in the walk-in unit where people could get medical attention for non-emergency care without needing an appointment. And I had the strangest, instantaneous reaction to an Asian man in his sixties who came in for treatment. As soon as I saw him, I was terrified. I began to sweat and my heart beat faster. The fight-or-flight response activated in me for no apparent reason.

I composed myself and examined the gentleman. But at the end of the evaluation, I confessed to him that I had a very strange feeling seemingly upon seeing him for the first time and I asked him if he might know why I had reacted this way. At that, the man smiled rather menacingly, made a gesture in which he pressed his thumb up and down on his fist, and declared, "Okay, Yank, now you die!" He went on to identify himself as the actor who had played the villain "Tokyo Joe" in early World War II movies I had seen as a child. He was later cast as the villain in Charlie Chan movies that I also remembered.

Without posing any actual threat to me, the appearance of this man in the walk-in unit had triggered the fight-or-flight response. My brain recalled the same patterns of nervousness I had experienced some twenty years before as a boy watching a frightening movie.

In other words, the things we recall or imagine are very real to the brain, as demonstrated by my colleague Dr. Stephen Kosslyn, a professor of psychology at Harvard.[6] Dr. Kosslyn performed an experiment in which people looked at a grid, a box of connected vertical and horizontal lines, with the letter A within the grid. He used a positron-emission tomography (PET) scan, a kind of brain imaging, to identify the areas where nerve cells fired when the people simply looked at the grid. Then he asked the same people to view a grid without the letter A embedded within it, but to "visualize" the letter A in their minds. Once again, a PET scan was taken to measure nerve-cell firing. Dr. Kosslyn found that the same areas of nerve cells were used. Whether actually looking at a grid with the letter A or imagining the grid with the letter A, the same nerve-cell firing patterns were stimulated in the brain.

So remembered wellness takes advantage of the fact that our brains cannot tell the difference between what is real and what we imagine or expect for ourselves. The opposite effect, called the "nocebo" effect, is also possible in which the worst that we imagine or predict for ourselves can come true, like a self-fulfilling prophecy. Some illnesses can be brought on or exacerbated and sometimes even death can result from a very potent experience with the nocebo effect.

The Way Emotions Are Assigned

Influential in both the nocebo effect and in remembered wellness is the discovery that emotions are critical to the proper working of our brains and bodies. From neurological research on people with some prefrontal lobe brain injuries, who experience a strange dispassionate approach to life after the injuries occur, we now understand that emotions help us assign priorities and weights to various aspects of our lives. Without emotions, decisions become very difficult to make because no one possibility has more meaning than another.

Medical research also has demonstrated that what humans observe or encounter in life is tagged with an emotion in the brain even before we have a chance to think about something or decide what we feel about it. Just as we do not always know what triggers a fight-or-flight response, our brains and bodies make split-second emotional decisions—an intuitive reaction that we are unused to thinking of as biological in nature. Psychologists, for example, have found that people presented with nonsense words or abstract shapes have instantaneous opinions about them. "We have yet to find something that the mind regards with complete impartiality, without at least a mild judgment of liking or disliking," says Dr. Jonathan Bargh, a New York University psychologist quoted in *The New York Times.*[7]

The Three-Legged Stool

Thus, remembered wellness is empowered by a much more complex sequence of brain/body functions than previously acknowledged, and it makes perfect sense that one's expectations of a medication, or one's beliefs in God applied to mental focusing, should contribute to its effects. With this in mind, I am working to change the way in which we practice medicine in the West, incorporating remembered wellness, the relaxation response, and other valuable methods of self-care. To me, the ideal medical model resembles a three-legged stool in which a patient's health and well-being are attended to with pharmaceuticals, surgery and other procedures, and self-care.

Faith and remembered wellness, namely the faith factor, obviously could be applied to all three legs of the stool. But my colleagues and I at the Mind/Body Medical Institute recommend the use of the faith factor as well as other self-care methods such as stress-reduction, nutrition, and exercise as a complement, not an alternative, to the traditional offerings of medicine. This is important because record numbers of Americans are turning to so-called alternative medicine, mostly, I believe, because patients ardently desire that medicine address spiritual as well as physical concerns and that caregivers afford them personal attention, treating them as individuals and not as disease states.

With the establishment of the National Institutes of Health Office of Alternative Medicine, unconventional therapies such as shark cartilage and Chinese herbs are for the first time being evaluated according to Western scientific standards. And even though I am sure that some of these therapies do help people, I predict that in the majority of cases, it is not because the techniques themselves have any inherent value, as patients may think. I do believe that chiropractic, massage, and other therapies that involve human touch eventually will prove to be inherently therapeutic. But when it comes to other unconventional methods, I am certain that belief in their effectiveness lends the therapies their power, just as belief lends traditional medicine much of its power.

As many as 60 to 90 percent of the medical problems that patients bring to their doctors' offices fall into the mind/body realm, those that can be alleviated by self-care methods including remembered wellness. Physicians used to appreciate this. In fact, up until the scientific age of the twentieth century, medicine was, in essence, remembered wellness. Before Dr. Louis Pasteur launched our modern understanding of "the germ theory," before Sir Alexander Fleming discovered penicillin, and before other great scientists discovered other medicines that saved lives, doctors had little or nothing other than placebos and reassurances to offer patients. The reputation of physicians was built on remembered wellness and on their ability to instill faith and confidence in their patients.

Then, as the germ theory encouraged us to do, scientific medicine began focusing, more and more precisely, on specific causes and solutions. We have almost entirely abandoned an appreciation for the context of health, namely life, and the beliefs that give it meaning. But as much as modern science and technology have changed the world, and as significantly longer as is our life span than previous generations would have dreamed possible, each precise, microscopic discovery we make still fails to answer the biggest question, How did life, as we know it, in all its intelligence and complexity, come to be? And if there truly was a master plan to this incredible and gigantic mystery of life, how large must the master be?

How Large Is God?

Whether or not science acknowledges "How large is God?" as a legitimate question, we continue to chip away at smaller mysteries that leave us baffled by the larger mystery. As close as we may come to "knowing" what compels a heart to pump or a mind/body to relax, we may never know the truth about the existence of God or God's influence.

Personally, I do believe in God and in an intelligent design to the universe. This is not a conviction I held before. It is, rather, the way that I have interpreted scientific evidence. It is inconceivable to me that without an underlying architect, life and intelligence could have come to exist. Thus, I believe in a world that is divinely influenced, as much as I believe evolution and biology are responsible for our physical make-up.

For my patients, however, the more important matter is not how we came to be designed but how we can use this design for our benefit.

I believe that humans are "wired for God." In every age and every civilization, humans have called upon a deity or deities. Indeed, faith quiets the anxiety of the mind and the body to an extent that no other belief can. Since we humans are the only species that recognizes its own mortality, the fact that we eventually will get sick and die produces enormous anxiety for us. And because the notion of God as an "infinite absolute" transcends death, faith in God can transcend the anxiety and pain that might otherwise consume us. Evoking remembered wellness in a powerful way, faith in God is a supremely therapeutic belief.

Some would argue that humans created this transcendent idea of God to bypass the painful reality of life and death having no greater meaning, and that this belief has become encoded in our genes. Others say that only God could have seen to it that humans believed in a transcendent power from the beginning, and that every generation since would evolve to crave faith, both physically and emotionally.

A Win-Win Situation

But whether or not there is a God, whether or not a force bigger than ourselves created us, and whether or not we were designed to pray and to be nourished physically by spirituality, faith in God, or in some form of the infinite absolute, is good for us. It does not matter to the health of the human body from whence this healthy spirituality comes, whether humans created the idea of God to soothe their anxious mortal souls or whether God wired us to be physically enhanced by believing in and actively worshiping God. Either way, we benefit; it is a win-win situation, although to some, this may sound like a cold assessment of religious faith.

I believe that there are other benefits to the human experience of faith. My patients frequently have reported experiencing spirituality in the elicitation of the relaxation response in the same rather mysterious way that people always have in quiet reflection and meditation. When people experience this spirituality, they cannot pin it down to anything more than "a presence of a power that is close to them."

In her book, *A History of God,* Karen Armstrong reiterates how important the God accessed through meditation and prayer has been to religious believers around the globe.[8] When humans abandon this notion and try to limit or overpersonalize God, Armstrong believes danger is in store. "He can be a mere idol carved in our own image, a projection of our limited needs, fears and desires. We can assume he loves what we love and hates what we hate, endorsing our prejudices instead of compelling us to transcend them. When he seems to fail to prevent a catastrophe or seems even to desire a tragedy, he can seem callous and cruel. . . . The very fact that, as a person, God has a gender is also limiting. . . ."

So indeed, we may never know how large God is. Nor does it work to our advantage to apply human standards to exploring God's influence. Maybe it is better for us if the Larger Mystery remains. It is, however, incumbent upon science to continue to study the role of God and faith in God. And we must respect religious believers and the role belief plays in their health and well-being.

My initial inklings that something was amiss in traditional medicine launched a long scientific quest, at the end of which I can say that faith is not only essential to our survival but too important for us to set aside. People benefit from faith in extensive and profound physical ways that medicine cannot afford to ignore.

As heartily as we may have tried, René Descartes and those of us scientists who followed his lead have failed to separate mind from body, biology from spirituality, and faith from facts. And despite what I believed thirty years ago, faith does yield facts. And so far, the evidence is compelling. Faith's influence over us is enormous, and for many, faith in God is the most powerful of all beliefs. We are just beginning to appreciate faith's wonderful implications for our health and our lives.

Notes

1. Herbert Benson with Marg Stark, *Timeless Healing: The Power and Biology of Belief* (New York: Scribner, 1996) pp. 1–350.
2. Henry K. Beecher, "The Powerful Placebo," *Journal of the American Medical Association* 159 (1995): 1602–1606.

3. Walter B. Cannon, *The Way of an Investigator: A Scientist's Experience in Medical Research* (New York: W. W. Norton, 1945).

4. Robert D. Orr and George Isaac, "Religious Variables Are Infrequently Reported in Clinical Research," *Family Medicine* 24 (1992): 602–606.

5. Jeffrey S. Levine and Preston L. Schiller, "Is There a Religious Factor in Health?" *Journal of Religion and Health* 26 (1987): 9–36.

6. Stephen Kosslyn, *Image and Brain: The Resolution of the Imagery Debate* (Cambridge, Mass.: MIT Press, 1994).

7. Jonathan Bargh as quoted in Daniel Goleman, "Brain May Tag all Perceptions with a Value," *New York Times,* August 8, 1995, p. C1.

8. Karen Armstrong, *A History of God: The 4,000-Year Quest of Judaism, Christianity, and Islam* (New York: Knopf, 1993).

No Place for a Small God

HOWARD J. VAN TILL

No Space in this Universe for God?

Much is being written these days, often in the name of modern science, about God's place in our spatially vast universe. As I reflect on some of this literature, however, I am struck by a remarkable irony. Only recently, within the twentieth century, have we learned how to measure the vastness of space in the physical universe. With optical and radio telescopes, we are now able to see galaxies and quasars at distances in the neighborhood of 15 billion light-years. Furthermore, from the red-shift in galactic spectra, we have learned that these astronomical giants are not standing idly in their cosmic remoteness, but they are receding from us at speeds roughly proportional to their distance.

We are, by definition, at the center of the *visible* universe—that portion of the universe that we are able to see. Even if there are galaxies beyond the 15 billion light-year horizon of visibility, we would be unable to see them because their light cannot reach us. Light emitted from galaxies receding at speeds beyond the speed of light is red-shifted to invisibility.

Even more difficult to comprehend than either the vastness of space or the recessional motion of galaxies is the idea that galactic recession comes about, not because galaxies are moving *through* space to even more distant locations in it, but rather because space itself is expanding. The distance to cosmologically remote galaxies is increasing because the amount of space between them and us is growing. The universe is expanding because the measure of space itself is increasing. Light from galaxies beyond the horizon of visibility is unable to reach us because the distance between them and us is growing at a rate greater than the speed of light. Clearly, our "common-sense" conceptual vocabulary regarding space and motion soon becomes inadequate to speak or think about such remarkable phenomena.

Because the universe is vast and growing ever larger, one might suggest, then, that there would now be more "room" for God than ever before. But both the reality and presence of God within the physical universe have come under severe question. In the minds of some persons, what scientific cosmology has discovered is not more space for God but, ironically, less space—perhaps no space at all.

The Myth of the Self-Creating Universe

The first, and very common, version of the no-place-for-God sentiment that I will consider in this essay is the simple presumption that the universe needs no Creator because it possesses the capacity for bringing itself into existence from nothing. Authors inclined toward scientism would add that it is within the competence of *science* to demonstrate that our universe needs no Creator as the source of its existence. In a brief book curiously titled, *The Creation*, British physical chemist Peter Atkins says, "My aim is to argue that the universe can come into existence without intervention, and that there is no *need* to invoke the idea of a Supreme Being in one of its numerous manifestations."[1] Elsewhere the author expresses his wish that the reader "admit that science is extraordinarily strong and if we disregard (as I argue we should) the question of a 'purpose' for the world, that it appears to be on the verge of explaining everything."[2] Furthermore, says Atkins, "the only way of explaining the creation is to show that the Creator had absolutely no job at all to do, and so might as well not have existed."[3] Toward the end of this book, reflecting on what he presumed to have accomplished, the author informs us that "we have been back to the time before time, and we have tracked the infinitely lazy Creator to his lair (he is, of course, not there)."[4]

Atkins repeats this no-need-for-God sentiment in a more recently published essay on "The Limitless Power of Science" in which his goal is to demonstrate that, although poets and theologians have, by his measure, contributed nothing of value to our self-understanding, "reductionist science is omnicompetent" to explain everything that is potentially explainable, even our very existence. It is clear that Atkins has his own all-inclusive definition of "science," and that it incorporates a host of naturalistic metaphysical propositions presented as if they were derived from what the rest of us ordinarily think of as natural science. "Science," he says, "is in the midst of showing that the concept of existence can survive the absolute simplicity of stripped down explanations and their ramifications." Thus, "While poetry titillates, theology obfuscates, science liberates."[5] An unusual concept of "liberation," I might suggest, if it entails the presumption that we and the universe of which we are a part possess no ultimate significance of the sort that both poets and theologians aspire to articulate.

Consistent with his earlier claims in *The Creation*, Atkins expresses in this essay his conviction that the goal of "science" (the label he continues to use for his comprehensive, scientifically informed, naturalistic worldview) should be far more than a mere description of the way the universe works. For Atkins, the ultimate and, in his judgment, achievable goal of science must be to explain cosmogenesis—the universe's coming into being. And in its quest for constructing this ultimate account of cosmic existence, this "science" must, he says, abide by high standards.

> [I]t must achieve all this by starting from something without a precursor; that is, it must start from nothing at all. The scientific account of cosmogenesis cannot stop when it has accounted for the universe springing from a seed the size of a Sun, nor when it has arrived at a seed the size of a pea. Nor can it stop at any smaller seed. A seed the size of a proton implies that that seed had to be manufactured, placed there by some cosmic pre-existing gardener. Science will be forced to admit defeat if it has to stop at a seed of any size. ... If we are to be honest, then we have to accept that science will be able to claim complete success only if it achieves what many might think impossible: accounting for the emergence of everything from absolutely nothing. Not almost nothing, not a subatomic dust-like speck, but absolutely nothing. Nothing at all. Not even empty space.[6]

Well, then, let us be honest. Can one reasonably presume that that which is properly named "absolutely nothing" nonetheless is gifted with the astounding capacities to transform its nothingness into something? Philosophers long ago offered the proposition, *ex nihilo nihil fit*, or *from nothing nothing comes*, and it remains as reasonable a proposition now, I believe, as it was in antiquity. As one philosopher expresses it, "Though dressed up in the guise of a scientific theory, the thesis at issue here is a philosophical one, namely, can something come out of nothing? ... The principle *ex nihilo nihil fit* seems to me to be a sort of metaphysical first principle, one of the most obvious truths we intuit when we reflect seriously."[7] One might well propose the eternal existence of a something that possesses the capacities to transform itself or to bring something else into being, but to propose that "absolutely nothing" produces, by any means, any form of something strikes me as a singularly vacuous proposal.

But Atkins asks for even more of his absolutely nothing. It must have the powers not merely to produce a something, but a something as remarkable as *this universe*. Many authors have called attention to the noteworthy list of "anthropic cosmological coincidences" that characterize the formational economy of this universe. (By the term "formational economy," I mean the set of all capabilities resident in physical or material systems that contribute to the self-organization or transformation of any inanimate or biotic form in the universe.) Without going into the details, which can be found in a number of excellent sources, the essential point is this: If, as appears to be the case, all of the life forms (including ourselves) now present in the universe have been actualized as the outcome of matter's capacities for self-organization and transformation, then a lengthy list of matter's particular properties and capabilities must be "just right."[8]

Those entries in the just right list that are of direct concern to scientific cosmology are called anthropic cosmological coincidences— *anthropic* because they are perceived to be prerequisites for the development of the human life form, *cosmological* because they affect the outcome of processes directly relevant to physical cosmology, and *coincidences* because there appears to be no material reason why they should be precisely as they are. One way in which these coincidences can be made to appear unexceptional is to posit the existence of a myriad of possible universes, of which ours happens to be one with an anthropically just right set of properties. Universes not having such a propitious set of properties will, of course, contain no observers to note the failure of their universe to make their existence a possibility. While such a proposition may constitute a clever logical strategy for avoiding the task of explaining these coincidences, one must wonder how it makes its move from a logical possibility to an intellectually satisfying explanation of why this universe is as remarkable as it is.

To get just a hint of the character of these coincidences, consider an example in nucleosynthesis. In order for carbon-based life to be possible, there must be atoms of carbon (and a host of other elements, of course) in the proper abundance. But the nuclei of carbon atoms are products of specific thermonuclear fusion reactions that occur within the interiors of stars as part of stellar evolution. By investigating the chain of fusion reactions that are responsible for the production of carbon in the proper abundance, it has become clear that several very specific properties of the nuclei of helium, beryllium, carbon, and oxygen had to be "just right." Forming nuclei of C^{12} directly from the collision of three nuclei of He^4 proceeds much too slowly to be of any value here. However, if the unstable nucleus of Be^8 were to have a lifetime long compared with the $He^4 + He^4$ collision time, then it could function as an intermediate step in the process and greatly increase the production of C^{12}. Anomalously, it does. However, for that autocatalytic reaction to take place, it must also be the case that the C^{12} nucleus has an energy level slightly above the rest energy of $Be^8 + He^4$ (7.3667 Mev). Remarkably, it does, at 7.6549 Mev. However, in order that the carbon so formed not be immediately transformed into oxygen by the addition of one more He^4 nucleus, it must also be the case that the O^{16} nucleus does *not* have an energy level just above the rest energy of $C^{12} + He^4$. Fortunately, it does not.

The bottom line is that a minute change in the value of any one of these energy levels or lifetimes would have made life as we know it physically impossible. Far more examples of the need for parameter values to fall within narrow limits could be cited. Hence the common expression that the fundamental properties of the universe appear to be "fine-tuned" to make our appearance possible. Is our universe the only viable one, or are there other possibilities based on entirely different parameter values or different forms of matter? In all honesty, we must say that we do not know the answer.

Instances of this cosmological fine-tuning can be found in phenomena at all levels of material behavior—from the submicroscopic world of elementary particles, in which the relative strengths of the four fundamental forces (gravitational, electromagnetic, strong nuclear, and weak nuclear) must lie within a very small range of what is observed to be the case, to the macroscopic world of galaxies receding from one another at the rate of spatial expansion, the exact value of which is crucial to the timescale of cosmic history.

In case after case, the more carefully we look at what appears to have taken place in the formational history of the universe of galaxies, stars, planets, and the atoms of which we are made, the more we become keenly aware of how dependent our presence is on the precise values of an extensive menu of fundamental physical parameters. Expanding the scope of our scrutiny to the biotic world and the formational history of life forms we would, I believe, become even more astounded at the encyclopedic list of "anthropic biological coincidences" regarding the properties and behavior of atoms and molecules that must have been satisfied in order for our appearance in the macroevolutionary drama.

In full awareness, I presume, of the fact that our presence is dependent upon the actuality of numerous anthropic coincidences—the few that we know about, along with the presumably far longer list of those that we have not yet discovered—Atkins nonetheless maintains his unfaltering faith in the competence of naturalism to provide an explanation of cosmogenesis from within the workings of nature itself. "That such a universe as ours did emerge with exactly the right blend of forces may have the flavor of a miracle, and therefore seem to require some form of intervention. But nothing intrinsically lacks an explanation. We cannot yet see quite far enough to decide which is the right [scientific] explanation, but we can be confident that intervention was not necessary."[9] Presuming that by "intervention" Atkins means any form of action by a being that transcends the physical universe, this is an extraordinarily bold (and, I would add, groundless) claim.

Where most thinkers, rightly I believe, see an unrealistic and unattainable task in this quest for comprehending the self-transformation of an absolutely nothing into a robust something, Atkins sees only an interesting challenge for scientific theorizing. "We shall, in a sense, need to model nothing, and to see if its consequences are this world. I don't regard this as impossible or ludicrous; I regard it as the next logical step for the development of the paradigms of science."[10] In contrast to Atkins, however, most of us *would* regard this enterprise as both impossible and ludicrous. Rhetorical bluster may intimidate those who are unsure of their case in a debate or in a court of law, but I have never known it to generate profound comprehension of substantive matters.

Atkins' faith in the omnicompetence of science is repeated in his book, *Creation Revisited*, with, it must be noted, extensive borrowing of rhetoric from the poetic language of the Bible.

> In the beginning there was nothing. Absolute void, not merely empty space. There was no space; nor was there time, for this was before time. The universe was without form and void.
>
> By chance there was a fluctuation, and a set of points, emerging from nothing and taking their existence from the pattern they formed, defined a time. The chance formation of a pattern resulted in the emergence of time from coalesced opposites, its emergence from nothing. From absolute nothing, absolutely without intervention, there came into being rudimentary existence. The emergence of the dust of points and their chance organization into time was the haphazard, unmotivated action that brought them into being.[11]

It would appear that in spite of his expression elsewhere of a low regard for the ability of poetry to express anything of value, he nonetheless freely employs the poetic genre to titillate his readers with his lyric expression of unbounded faith in the ability of absolute nothingness to self-transform into our robustly functioning universe. Atkins' self-creating universe must, however, like all other self-creating universes, begin with some form of self. Thus, what Atkins has done is not to demonstrate that a Creator is unnecessary but merely to give his own version of a Creator a succession of new names—first *absolutely nothing*, then *this universe*. In comparison to his claims, Atkins' accomplishment seems quite modest.

Atkins may be among the most strident advocates of the metaphysical doctrine that the universe needs no Creator, but he is not alone in giving such a conjecture serious consideration. In his introduction to Stephen Hawking's immensely popular book, *A Brief History of Time*, astronomer Carl Sagan commends both the book and its author for providing readers with "lucid revelations on the frontiers of physics, astronomy, cosmology, and courage." He then goes on to say,

> This is also a book about God ... or perhaps about the absence of God. The word God fills these pages. Hawking embarks on a quest to answer Einstein's famous question about whether God had any choice in creating the universe. Hawking is attempting, as he explicitly states, to understand the mind of God. And this makes all the more unexpected the conclusion to the effort, at least so far: a universe with no edge in space, no beginning or end in time, and nothing for a Creator to do.[12]

Hawking does raise the question concerning the universe's need for a Creator, but the ambivalence of his answer is somewhat puzzling. In Hawking's own words:

> The idea that space and time may form a closed surface without boundary also has profound implications for the role of God in the affairs of the universe. With the success of scientific theories in describing events, most people have come to believe that God allows the universe to evolve according to a set of laws and does not intervene in the universe to break these laws. However the laws do not tell us what the universe should have looked like when it started—it would still be up to God to wind up the clockwork and choose how to start it off. So long as the universe had a beginning, we could suppose it had a Creator. But if the

universe is really self-contained, having no boundary or edge, it would have neither beginning nor end: it would simply be. What place, then, for a Creator?[13]

The implied answer to Hawking's question is, as Sagan seems to have presumed in his introduction, "No place at all." I must, however, entertain the possibility that Hawking is not necessarily insisting on that negative answer, but he is perhaps challenging the theists among his readers to articulate some more profound and positive answer. After reflecting on Hawking's conjecture regarding a self-contained universe that had neither beginning nor end, an interviewer once asked him, "Does that mean there was no act of creation and therefore that there's no place for God?" Hawking's reply was, "I still believe the universe has a beginning in real time, at a big bang. But there's another kind of time, imaginary time, at right angles to real time, in which the universe has no beginning or end. This would mean that the way the universe began would be determined by the laws of physics. One wouldn't have to say that God chose to set the universe going in some arbitrary way that we couldn't understand. It says nothing about whether or not God exists—just that He is not arbitrary."

"But," the interviewer persisted, "I think that many people do feel you have effectively dispensed with God. Are you denying that then?" To this direct question Hawking responded, "All that my work has shown is that you don't have to say that the way the universe began was the personal whim of God. But you still have the question: Why does the universe bother to exist? If you like, you can define God to be the answer to that question."[14] In another context, Hawking's response to the same question was, "I don't know the answer to that."[15]

There are substantive differences among the views expressed by Atkins, Sagan, and Hawking that must be respected. Nonetheless, they do deal with a common and very fundamental question: Is a Creator necessary to give being to a universe? Hawking, with a blend of both tentativeness and ambiguity, appears willing to keep the question open for thoughtful reflection. Atkins, on the other hand, presumes to have made such a Creator unnecessary simply by postulating, without a hint of tentativeness, that absolutely nothing has the capacity to transform its nothingness into something like this universe. Surely this represents a singularly facile way do deal with a substantive question— solve the mystery of existence by sheer assertion alone, simply declaring that the universe needs no transcendent power to give it being.

The Robust Formational Economy Principle

A second version of the no-place-for-God thesis places less emphasis on questions regarding either the temporal beginning or the ultimate source of the universe's existence and it focuses instead on the robustness of its powers for organizing itself into novel or more complex forms. The natural sciences, in their activity of constructing theories regarding the formational history of both inanimate structures (galaxies, stars, and planets, for instance) and living organisms, do ordinarily presume that matter and material systems possess a sufficiently robust set of self-organizational capabilities to make possible the actualization in time of all the physical and biotic forms that have ever appeared in the history of the universe. This "robust formational economy principle," as I would name it, is seldom given explicit statement. Nonetheless, it constitutes, I believe, one of the foundational presuppositions of contemporary science, especially in its theorizing regarding formational histories.

In the context of this essay, the question at issue is, Does this principle provide any support for the no-place-for-God thesis? The rhetoric of many proponents of a naturalistic world view is built on the presumption of a "yes" answer. My answer is, on the contrary, "no." Not only does the robust formational economy principle fail to provide warrant for the no-place-for-God thesis but it makes the intentional action of a Creator/Provider even more essential. Realizing that my thesis runs counter to the more familiar rhetoric, I need to develop my case with some care.

First, what circumstance might encourage a proponent of naturalism to believe that his or her world view would gain support if the robust formational economy principle were true? Ironically, that encouragement may in large part arise from the common belief within the Christian community, particularly within its conservative evangelical portion, that the theological doctrine of creation either entails or is strengthened by the *special creationist* picture of God's creative activity. A major portion of the North American Christian community, for instance, posits not only that God gave being to the elemental substances of the universe at the beginning, and that God continually sustains the Creation in being, but also that the arrangement of those elemental substances into specific physical and biotic forms required a succession of "special" divine creative acts in the course of time. By a "special" creative act, we here mean any immediate divine action performed for the purpose of bringing a new structure or life form into being, either by *creatio ex nihilo* or by coercing extant matter into configurations that it would not have been able to achieve by employing only its own limited powers for self-organization or transformation.

This special creationist position comes in numerous versions: *young-earth special creationism* (which holds that the Bible specifies a 6,000-10,000-year chronology); *old-earth special creationism* (which grants the 15 billion-year chronology favored by the natural sciences); and some variants of both *progressive creationism*[16] and the resurgent *intelligent design* perspective.[17] Common to all these versions of special creationism, however, is the idea that the present array of physical and biotic forms could not have been assembled without a number of special creative acts—often referred to as "divine interventions"— in the course of time.

A fundamental presumption of special creationism (although it is not always stated as explicitly as this) is that the self-organizational and transformational capabilities of atoms, molecules, and cells are not sufficient to bring about the actualization of all the structures and forms that we now see, at least not all the biotic forms. Thus it must be the case, goes the familiar argument, that God has "intervened" at certain times to introduce specific novel forms that could not have arisen by the exercise of any "natural" capabilities.

With all due respect for those proponents of special creationism whose intent is to maintain a high regard for divine power over the created world and to ensure a place of need for the immediate exercise

of that power to actualize at least some biotic forms, I must nevertheless state candidly that, even as one wholly committed to the Christian
faith, I have substantive difficulties with this perspective. Contrary to
the beliefs of many of its contemporary proponents, I do not find warrant for the claim that the Scriptures require one to adopt this perspective.[18] Neither do I find any merit in the common presumption that
the special creation scenario constitutes one of the "deliverances of the
saints," that is, one of the foundational beliefs of the early Christian
community.[19] Furthermore, I would argue that the special creationist
position invites the *naturalistic challenge*—a form of the no-place-for-
God sentiment—that can be paraphrased as, *If there are no gaps in the
formational economy of the universe, what need, then, for a Creator?*

There is no scarcity of literature that conveys this naturalistic challenge. Atkins, cited earlier, provides an abundance of examples. Zool-
ogist Richard Dawkins, in *The Blind Watchmaker* and other works,
presumes to have established the no-place-for-God thesis once and for
all.[20] A similar claim is offered by Francis Crick in *The Astonishing Hypothesis.*[21] For the moment, however, let me evaluate and respond to
the line of thought developed in a recent book by philosopher Daniel
C. Dennett, *Darwin's Dangerous Idea.*[22]

One of the questions that Dennett considers is, How do the things
we see in the world around us, especially living things, come to exhibit
design, whether in their internal workings or in their adaptation to the
peculiarities of some environment? For Dennett, the term "design"
functions as the generic category for any feature of the world, especially of its life forms, that is likely to give one the impression of being
the outcome of an intentional action of an intelligent agent. More
than mere order, "Design is Aristotle's *telos,* an exploitation of Order
for a purpose, such as we see in a cleverly designed artifact."[23]

The first of two major possibilities that Dennett considers is a traditional "mind-first" approach—things in our world manifest design because they proceed from a pre-existing mind—an approach that Dennett associates with John Locke.

> [I]f Locke is right, Mind must come first—or at least tied for
> first. It could not come into existence at some later date, as an ef-

fect of some confluence of more modest, mindless phenomena.
. . . The traditional idea that God is a rational, thinking agent, a
Designer and Builder of the world, is here given the highest
stamp of scientific approval: like a mathematical theorem, its denial is supposedly impossible to conceive. . . . And so it seemed to
many brilliant and skeptical thinkers before Darwin.[24]

Dennett sees a similar sentiment in the phenomenon of eighteenth
century natural theology, with its apologetic strategy of arguing from
the empirical observation of design to the conclusion of the existence
of a designer.

> The overwhelming favorite among purportedly scientific argu-
> ments for religious conclusions, then and now, was one version
> or another of the Argument from Design: among the effects we
> can objectively observe in the world, there are many that are not
> (cannot be, for various reasons) mere accidents; they must have
> been designed to be as they are, and there cannot be design with-
> out a Designer; therefore, a Designer, God, must exist (or have
> existed), as the source of all these wonderful effects.[25]

It is, I believe, important to note here that the concept of designer,
as it was employed in the eighteenth century by clergyman William
Paley and others, was based on the artisan metaphor. One person, the
artisan, did both the planning and the fabrication of what was
planned. Paley's watchmaker, for instance, did both the planning and
the construction of the watch. Paley's designer (like his watchmaker)
had both a mind (to plan or intend) and the divine equivalent of
"hands" (the power to manipulate raw materials into the intended
form). The concept of special creation clearly resides in the same conceptual territory as that of an artisan/designer.

As a proposed answer to the question, Whence the appearance of
design? Dennett forcefully rejects the "handicrafter-God" of both special creationism and the argument from design. He characterizes such
approaches as ill-conceived attempts to inject supernatural explanations into circumstances where natural explanations would suffice. In
Dennett's colorful metaphor, he sees no need to appeal to a "sky-

hook" (the top-down action of some higher power) when a "crane" (the bottom-up action of some extant natural mechanism) is able to do the job of lifting the biotic system to new heights of configurational complexity.

> The skyhook concept is perhaps a descendant of the *deus ex machina* of ancient Greek dramaturgy: when second-rate playwrights found their plots leading their heroes into inescapable difficulties, they were often tempted to crank down a god onto the scene, like Superman, to save the situation supernaturally. . . . [A] *skyhook* is a "mind-first" force or power or process, an exception to the principle that all design, and apparent design, is ultimately the result of mindless, motiveless mechanicity. *A crane*, in contrast, is a subprocess or special feature of a design process that can be demonstrated to permit the local speeding up of the basic, slow process of natural selection, *and* that can be demonstrated to be itself the predictable (or retrospectively explicable) product of the basic process.[26]

Dennett's position regarding the success of the concept of biological evolution is comprised of three fundamental claims at quite different levels (although his rhetoric does not demonstrate an awareness of these differing levels): The credibility of unbroken genealogical continuity among all life forms has, he says, been established; the concept of special creation has been, once and for all time, discredited; hence, the existence of the entire universe, complete with its remarkably robust formational economy, may be taken for granted as a starting point that needs no explanation.

The first claim is stated by Dennett with characteristic immodesty: "To put it bluntly, but fairly, anyone today who doubts that the variety of life on this planet was produced by a process of evolution is simply ignorant—inexcusably ignorant, in a world where three out of four people have learned to read and write."[27] Suppose, however, we were to restate that claim more politely and in a more comprehensive form, drawing on some of the conceptual vocabulary introduced above: *We have substantial empirical warrant for presuming that matter and material systems do possess the resident capabilities for self-organization and transformation of the sort envisioned by evolutionary theorizing in sciences such as cosmology and biology.* The concept of evolutionary continuity employed in contemporary scientific reconstructions of the formational history of both physical and biotic forms is strongly supported by its fruitfulness in accounting for a plethora of observable features of the physical and biotic world. Stated in this form, I would be not only willing but eager to give my assent. Although I cannot, of course, prove its validity, I do judge it to be a warranted belief.

Dennett's second claim follows quite directly from the first. Given the credibility of unbroken genealogical continuity, we are then warranted in rejecting the concept of special creation—a picture of divine creative action centered on the presumption that matter and material systems do *not* possess sufficient self-organizational and transformational capacities to make evolutionary continuity possible, and that substantial gaps in the formational economy of the universe have been bridged by a succession of extraordinary divine creative acts in the course of time. Once again, I must say that I concur with Dennett in his rejection of the special creationist picture of the formational history of life forms.

The Greater Necessity of Mind

Now, having granted two of the three claims offered by Dennett, have I thereby "given away the store"? Have I, by my acceptance of evolutionary continuity and by my rejection of special creationism, made the acceptance of Dennett's naturalistic worldview the inevitable conclusion? The usual rhetoric of the popular creation-evolution debate would, I presume, say that I have indeed done so. But let us look more closely at that third claim. In essence it says that if we are warranted in rejecting the concept of special creation, then we are equally warranted in rejecting any concept of a Creator, including one whose comprehensive act of creation consists in the *giving of being* to a universe generously gifted with a formational economy sufficiently robust to make possible the evolutionary development of self-organizing physical and biotic systems.

In other words, Dennett's third claim is: If he has been able to discredit the particular concept of special creation, then he is fully warranted in rejecting the broad theistic doctrine of creation and in presuming that the existence of the entire universe, complete with its remarkably robust formational economy, may be taken for granted. But that is, quite clearly, a non sequitur of colossal proportions! One might even be tempted to say that anyone who fails to recognize that as a non sequitur is "simply ignorant—inexcusably ignorant, in a world where three out of four people have learned to read and write."

All that Dennett, or Dawkins, or other proponents of evolutionary naturalism have been able to demonstrate is that one particular scenario for the historical manifestation of God's creative work—the special creationist scenario—fails to gain empirical support. Contrary to the presumptions of special creationism, there is strong empirical encouragement for the expectation that the formational economy of the universe is sufficiently robust to account for the actualization of all physical and biotic forms known. Now, although that is a relatively simple proposition to state, it represents a truly astounding state of affairs!

Those persons who have some familiarity with the capabilities that atoms and molecules and cells must have in order for evolutionary development to be possible will recognize the profound significance of granting the credibility of the robust formational economy principle. The list of requisite capabilities is more extensive than our minds are capable of comprehending. For that matter, the list of requisite capacities that make possible just one day of our existence is beyond full comprehension.

How is it, then, that such a remarkable universe has being? Not only must we account for the existence of something in place of nothing but also of a something possessing a host of truly astounding capabilities. Here is, I believe, where the "mind-first" thesis of theism is in a position to offer an answer vastly more reasonable and satisfying than the "no-mind" thesis of naturalism. How does something (a universe, say) come not only to exist but also to possess a formational economy as robust as the one exhibited by our universe, as the natural sciences are just beginning to realize? If not as an intentional and generous gift from a Creator, then how? By consequence of the self-transformational powers of absolutely nothing? Atkins and Dennett would have us think so, but on what basis? On the convincing power of a boldly stated assertion or by some unaccountably propitious stroke of luck?

Dennett's presentation of "Darwin's dangerous idea" is crafted to give the appearance of doing away with the need for the prior existence of mind. But I must say that I find the rhetoric wholly unconvincing. In fact, I would argue from his premise (that special creationism has been discredited by the growing credibility of the robust formational economy principle) to precisely the opposite conclusion that a creative Mind is absolutely essential.

Dennett places much weight on the explanatory power of an *algorithmic* process—any material process whose outcome, no matter how complex in appearance, proceeds from the actions of basic material units (atoms, molecules, cells) behaving in accordance with relatively simple rules.

> Here, then, is Darwin's dangerous idea: the algorithmic level is the level that best accounts for the speed of the antelope, the wing of the eagle, the shape of the orchid, the diversity of species, and all the other occasions for wonder in the world of nature. . . No matter how impressive the products of an algorithm, the underlying process always consists of nothing but a set of individually mindless steps succeeding each other without the help of any intelligent supervision. . .[28]

But does the existence of a robust economy of algorithmic processes actually make mind unnecessary? Not at all. Suppose we were to grant that the relevant basic material units do in fact possess the capabilities to act out a set of algorithmic processes, and that the outcome might well be the self-organization of atoms to form molecules, molecules to form cells, cells to form organisms, etc. What thereby becomes unnecessary is not the action of a mind to intend or plan the elements of this robust formational economy, but rather the manipulative intervention of a "hand" to effect an act of special creation. What still requires mind is the conception of a sufficiently robust algorithmic economy. In fact, the more robust the requisite economy of creaturely capabilities, the *more*, not less, a creator/mind becomes absolutely necessary. The correct conclusion of Dennett's appeal to the existence of a robust economy of algorithmic processes is not that a mindful Creator is unnecessary, but rather that the creativity of that mind is far more extensive than has ordinarily been presumed and that the mindfully intended universe to which the Creator has given being is even more generously gifted with formational capabilities than we had initially realized.

Why is it, then, that Dennet and other vocal proponents of naturalism are inclined toward the no-need-for-God conclusion? Why would one be led to think that the fruitful functioning of algorithmic processes would displace a mindful Creator? Could it be that the recent popularity of special creationism and its handicrafter-God has, contrary to the good intentions of its proponents, provided the context for this misunderstanding? I believe a strong case could be made for this thesis. Dennett's own rhetoric points us in that direction.

> The resistance [to the concept of the continuity and sufficiency of material processes] comes from those who think there must be some discontinuities somewhere, some skyhooks, or moments of Special Creation, or some other sort of miracles, between the prokaryotes and the finest treasures in our libraries.[29]
>
> For over a century, skeptics have been trying to find a proof that Darwin's idea just can't work, at least not *all the way*. They have been hoping for, hunting for, praying for skyhooks, as exceptions to what they see as the bleak vision of Darwin's algorithm churning away. And time and again, they have come up with truly interesting challenges—leaps and gaps and other marvels that do seem, at first, to need skyhooks. But then along have come the cranes, discovered in many cases by the very skeptics who were hoping to find a skyhook.[30]

The problem lies, I would suggest, with the special creationist proposition of the necessity for skyhooks to preserve a place of need for God in the universe. When the argumentation for belief in a Creator becomes tightly coupled with the thesis that divine acts of special creation are required to bridge gaps in the formational economy of the universe, the proponents of naturalism are offered an easy apologetic strategy: demonstrate empirical support for the existence of a formational economy sufficiently robust to make macroevolution possible, and the handicrafter-God of special creationism loses his place in the universe. And if the handicrafter-God of special creationism is displaced, then why not boldly extrapolate to the presumption that all creator-gods have been displaced, including the God who is the ultimate giver of being?

Why not? Stated as simply as I am able, because the robustly equipped universe requires a giver of being just as much—even more, I would say—than does an incomplete universe with gaps in its formational economy. To give being to a universe as richly gifted as our universe appears to be requires a "large" God—large in creativity, large in generosity, and large in visionary anticipation of potentialities. The universe in which we live is no place for a small God—no place, for instance, for a God whose creative actions are confined to the bridging of gaps in an inadequate formational economy.

Proponents of special creation do not, of course, intend to hold a diminished view of their Creator. Nonetheless, there is a problematic feature of that perspective that could, and often does, lead critics of theism to restrict their picture of God's creative work to occasional episodes of "patchwork repair jobs." That may sound harsh, but let me explain. Recall that the special creationist perspective espouses a concept of divine creative action that includes, as an essential element, the provision that some specific physical structures or life forms must have been actualized, not by the exercise of creaturely capacities but by direct divine action (or "miraculous divine interventions") in the course of time. Furthermore, this special divine action is thought to be essential for the reason that creaturely capacities are deemed inadequate to accomplish the actualization of certain physical or biotic forms.

But why would it be the case that creaturely capacities are inadequate for the task of actualizing certain specific physical or biotic forms? Recalling that in the creationist perspective all creaturely capacities are God-given, we are bound to conclude that any instance of inadequate, or "missing," creaturely capacities is itself a product of divine intention. In other words, *the special creationist perspective necessarily entails the proposition that God intentionally withheld from the Creation a carefully selected set of creaturely capacities so that there would be gaps in its formational economy.*

Believing this to be the case, many proponents of special creation are actively engaged in seeking empirical evidence for the presence of these gaps in the formational economy of the universe. Thus, the special creationist perspective has the counterproductive effect of focusing one's attention, not on the rich array of remarkable gifts that do characterize the Creation's formational economy, but on those few that are presumed to have been withheld from it. Furthermore, if the mark of God's creative action is thought to reside primarily in instances of special creation or of "miraculous divine intervention," then there is a strong tendency to treat gaps in our human understanding of physical or biotic processes as evidence for the presence of corresponding gaps in the formational economy of the Creation. In its most facile form, the argument is, If I cannot conceive of a creaturely process to accomplish X, then X must have been accomplished by an act of special creation. In some presentations of this perspective, the term *special creation* is replaced by *intelligent design.*[31]

It is this sort of sentiment that invites zealous proponents of naturalism like Atkins, Dawkins, or Dennett to respond by saying, in effect, But there is growing empirical evidence that the universe *does* have the requisite capacities for self-organization and transformation to accomplish the formation of all of the physical structures and living creatures that have ever existed. This being so, then there is no further need for your handicrafter-God to bridge gaps in the universe's formational economy. Thus, there is no longer any room for your God in this universe.

Herein lies the tragedy of the creation-evolution debate and its inverted scoring system. As the debate is now structured and most commonly argued, each scientific discovery of some remarkable creaturely capacity that increases the credibility of evolutionary continuity is credited to the world view of antitheistic naturalism. The credibility of the Christian theism's historic doctrine of creation, on the other hand, is presumed to be affirmed primarily by empirical evidence for the existence of gaps in the formational economy of the universe. It is no wonder at all that the proponents of naturalism like this scoring system. The astounding thing to me is that so many persons within my own Christian community have come to accept it as well.

An Invitation and a Challenge

Let me close with an invitation to theists and a challenge to proponents of naturalism. I invite my fellow theists, particularly those who continue to hold to the special creationist picture of God's creative activity, to enlarge their portrait of the Creator. Allow that portrait to be large enough to include the expectation that the universe to which this Creator has given being has been generously gifted from the outset with a formational economy sufficiently robust to make possible its self-organization into the full array of physical structures and biotic forms that have ever been actualized. Develop the habit of perceiving the self-organizational capabilities of atoms and molecules and cells as evidence, not that God is unnecessary but that God has been unfathomably generous in the giving of being to his creation. Develop the habit of perceiving the fecundity of the Creation's formational economy as evidence, not that God need not be active but that God's faithful and essential action of blessing the efforts of his creatures has been fruitful beyond our powers to imagine.

Finally, I challenge proponents of naturalism, particularly those whose apologetic strategy begins with the employment of the natural sciences to discredit the handicrafter-God of special creationism, to face the reality that there are much larger portraits of God that now must be engaged. Recognize that neither the existence of the universe, nor the robustness of its formational economy, nor the fecundity of its developmental history are self-explanatory. Recognize that a facile extrapolation from the discrediting of special creationism to the denial of the need for a Creator to give being to the universe represents no substantive accomplishment. Allow the familiar proposition, *ex nihilo nihil fit*, to provide the occasion for asking, From what source, then, does our being proceed? How does the something that exists—this universe—come to have a formational economy as remarkably robust as we are just now beginning to realize? For what purpose did our source of being grant us the gifts that we possess? How then shall we employ these gifts?

Notes

1. P. W. Atkins, *The Creation* (San Francisco: W.H. Freeman & Company, 1981) p. vii.
2. Ibid., p. viii.
3. Ibid., p. 17.
4. Ibid., p. 115
5. P. W. Atkins. "The Limitless Power of Science," in John Cornwell, ed., *Nature's Imagination* (New York: Oxford University Press, 1995) pp. 122–32.
6. Ibid., p. 131.
7. William Lane Craig. "The Caused Beginning of the Universe," *British Journal for the Philosophy of Science* 44 (1993): 623–639.
8. For book length works, see John D. Barrow and Frank J. Tipler, *The Anthropic Cosmological Principle* (New York: Oxford University Press, 1986) and John Leslie, *Universes* (New York: Routledge, 1989). For a concise essay accompanied by a helpful set of references, see "Anthropic Principle," in Norriss S. Hetherington, ed., *Encyclopedia of Cosmology* (New York: Garland Publishing, 1993).
9. Atkins, *The Creation*, pp. 123–25.
10. Atkins, *Limitless Power*, p. 132.
11. P. W. Atkins, *Creation Revisited* (New York: W. H. Freeman, 1992) p. 149.
12. Carl Sagan, "Introduction," in Stephen Hawking, *A Brief History of Time* (New York: Bantam, 1988) p. x.
13. Hawking, *Brief History*, pp. 140–141.
14. Stephen Hawking, *Black Holes and Baby Universes and Other Essays* (New York: Bantam, 1993) pp. 172–173.
15. Ibid., p. 99.
16. See the classic exposition of this view in Bernard Ramm, *The Christian View of Science and Scripture* (Grand Rapids, Mich.: Eerdmans, 1956).
17. See, for instance, *The Creation Hypothesis: Scientific Evidence for an Intelligent Designer,* J. P. Moreland, ed., (Downers Grove, Ill.: InterVarsity Press, 1994). Also see Michael J. Behe, *Darwin's Black Box: The Biochemical Challenge to Evolution* (New York: The Free Press, 1996).

18. An indication of my own approach to the Scriptures on this issue can be found in my book, *The Fourth Day: What the Bible and the Heavens Are Telling us about the Creation,* (Grand Rapids, Mich.: Eerdmans, 1986).

19. An evaluation of this claim is the subject of my essay, "Basil, Augustine, and the Doctrine of Creation's Functional Integrity," *Science and Christian Belief,* April, 1996, pp. 21–38.

20. *Richard Dawkins, The Blind Watchmaker* (New York: W. W. Norton, 1986).

21. Francis Crick, *The Astonishing Hypothesis* (New York: Charles Scribner's Sons, 1994).

22. Daniel C. Dennett, *Darwin's Dangerous Idea: Evolution and the Meanings of Life* (New York: Touchstone, 1995).

23. Ibid., p. 64.

24. Ibid., p. 28.

25. Ibid.

26. Ibid., pp. 74, 76.

27. Ibid., p. 46.

28. Ibid., p. 59.

29. Ibid., p. 136.

30. Ibid., p. 75–76.

31. See note 17.

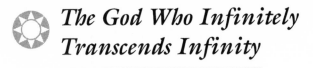

The God Who Infinitely Transcends Infinity

ROBERT J. RUSSELL

God-Talk: Glimmers of Knowing within the Horizons of Unknowing

Amidst the trappings of daily life, we occasionally sense the presence of something that, although lying beyond the limits of our ordinary world, reaches up from the infinite depths that sustain our world to touch our lives and fill us with hope and compassion. In these moments, we know the staggering beyond in our midst, the astonishingly tender caress of that caring ultimacy in which "we live, and move, and have our being"(Acts 17:28). We come away from these experiences enriched by a glimmer of understanding of what, or better yet Who, touched us, though our understanding lies within the all-encompassing mystery of its infinite, unknown source.

In seeking to articulate this numinous experience, we use the sacred and sublime word that millions of Jews, Christians, and Moslems have used for centuries: God. With this word, we affirm the personal existence and ultimate reality of that which transcends us utterly, a transforming power and singular presence that has touched the innermost reaches of our self and that even we can barely grasp. Our experience has left us staggered with the realization, given as an irreducible datum, that this God who is absolute mystery is also our Creator and Redeemer, the source and goal not only of our life but also of the existence and destiny of the cosmos.

It seems entirely fitting then that our talk about God should start with how little we know about God. Theology, which strictly speaking is "God-talk" (from the Greek word *theos* for God, and *logos* for

137

word), starts with what we do not understand before it seeks to say something about that which we understand—at least in part. In the long traditions of Western monotheism, the way of unknowing leads us, and the way of knowing follows behind. These ways have come to be called the *via negativa* (the negative way of denial) and the *via positiva* (the positive way of affirmation), respectively. These ways are important when we talk about grace, creation, or atonement, for example, but they are particularly important when we attempt "God-talk" in the particular sense of discourse about the nature of God.

This means, in turn, that when we talk about God, we should begin by attempting to say something about those divine attributes that we only know by negation, that is, by their utter contrast with our experience of ourselves and our universe. Such statements are called "apophatic" meaning "to deny." The most inclusive of these attributes is the incomprehensibility of God. The word "God" stands for a self-surpassing mystery, an ineffable reality lying beyond all our thoughts and sensations, that which is wholly other. In a similar way, we speak apophatically when we refer to God's infinity and eternity as being in contrast with our finite and mortal world. Everything we experience in this visible world seems finite and limited, temporal and transitory. It comes into existence, endures for a while, then passes away. It is really here now, yet soon it is gone forever. By utter contrast, however, God alone is invisible, unlimited and infinite, everlasting and eternal, unchanging and constant, the ground of being, the ultimate, unseen but ever-present reality. But that contrast, too, tells us something inestimably important about God's relation to the world: It is only by being incomprehensibly different from this world that God can be the source of this world and its final home. Thus the way of unknowing is, paradoxically, a way of knowing and a source of confident hope.

Having said that, we also attempt to speak about what God has revealed to us about God's nature and purposes, through scripture, tradition, and personal experience—and now in the past three centuries and increasingly today through the great discoveries of the natural sciences. Here we can talk about God not by sheer contrast but by making a positive analogy with key experiences in our own lives and with the world in which we live. Such statements are called kataphatic since they involve an analogy between God and what we know about in the world.

When we experience wonder and awe at the immensity and beauty of the universe, we are led to think of God as utterly wondrous, terribly awesome, and the source of ecstatic beauty. When we experience love in our lives; when we are forgiven our iniquities by those we have wronged; when we know the goodness of home, hearth, health, and family; then we speak of God as perfect love, unconditional mercy, the source of all that is good, and our final home beyond death and grief. Most important, when we look up from the routine of life and witness the sacred in our midst, as Moses did when he turned aside from tending to his sheep to go to see the burning bush (Exodus 3:1-6), then we confess God as utterly holy.

Thus, we are surrounded with the mystery of God that surpasses all knowing; yet we know that this God seeks us and would be known by us, and we move ahead in the light of this knowledge. To do so, we must remember the poverty of faith in light of the surpassing mystery of God, and cloak all that we wish to affirm in the spreading folds of our unknowing. We explore the kataphatic on the grounds of the apophatic. The way of faith is always the way of humility.

Remarkably, many scientists take a similar approach to the mystery of the universe around us and in us. We find that the exploration of the universe requires that we affirm and celebrate what we have discovered only within the shadow of that greater awareness of all that we do not know and perhaps never will.* Near the end of his life, Sir Isaac Newton, one of the greatest scientists of all times, was to write:

> I do not know what I may appear to the world; but to myself I seem to have been only like a boy playing on the sea-shore, and diverting myself in now and then finding a smoother pebble or a

---

*We can put this more formally by claiming that the scientific method, much like the theological method, proceeds by hypotheses held tentatively, intended as true, tested by data, and subject to acceptance or rejection by the consensus of the community. All genuine knowledge—religious or scientific—is finally a small light shining in a vast darkness, a known illuminating the unknown.

prettier shell than ordinary, whilst the great ocean of truth lay all undiscovered before me. [1]

And so the paths of science and theology both reflect the way of knowing as surrounded and held in perspective by the way of unknowing.

God-Talk: Focus on Infinity

I want to explore here the theological conversation about God and the scientific conversation about the world to see how our recent advances in cosmology and mathematics bear on this conversation. The key concept here will be infinity, including traditional notions of God's infinity in space and God's infinity in time, or eternity. Now infinity is not a biblical term, nor does it carry a single meaning in the theological, philosophical, or scientific communities.[2] Moreover, whenever we move from the language of one community to that of another, and across periods of time and history, we must recognize the differences, as well as the similarities, of language.

Still, for our purposes, I assume that sufficient continuities exist between these communities and periods of time to allow us to compare the meaning of the term infinity in a reasonable way. I hope to show that this term has, in fact, played a key role in all three communities: scientific, philosophical, and theological. Because of this, recent discoveries in the scientific community about infinity can have a direct bearing on our understanding of the concept, God.

Why choose infinity? As I have suggested, when we say that God is infinite and eternal, we normally do so by the *via negativa:* by way of utter contrast with the finite and temporal world that God created. This reflects a view of infinity that dates back to the ancient Greeks, where infinity is defined apophatically through its contrast with the finite: The infinite is called *apeiron,* meaning unbounded, unlimited, or formless. But is infinity utterly beyond our understanding, or can we know something about the meaning of infinity without exhausting all that can be true of it, reflecting the kataphatic way of knowing?

Recent developments in mathematics and in scientific cosmology may shed provocative new light on this issue. First, mathematicians have now given us a new conception of infinity that is much more complex than previously thought, with layers of infinity leading out endlessly to an unreachable Absolute. In effect, we now can say a lot more about infinity than merely that it contrasts with the finite—indeed, an infinite amount more! How might these revolutionary discoveries in mathematics enhance our concept of God as infinite?

Second, we have moved from the ancient cosmologies of the Bible and the Greeks into the immensity of big bang cosmology, in which the universe is billions of light-years—or perhaps even infinite—in size and expanding in time. How can this revolutionary understanding of the immensity of nature help us appreciate the immensity and ineffability of the God who is its Creator?

Some see these developments as leading to pantheism, the equating of God and the universe, or to atheism, the outright rejection of God. Clearly these questions are extraordinarily complex and the relevant literature is enormous. Still, I suggest that these fascinating developments in mathematics and cosmology actually clarify and increase our understanding of God the Creator as infinite and eternal and of the universe as dependent on God for its existence and purpose. [3]

Brief History of Infinity in Mathematics and Cosmology [4]

> It is possible to regard the history of the foundations of mathematics as a progressive enlarging of the mathematical universe to include more and more infinities. The concept of infinity in Greek and medieval thought includes a discussion first of philosophy and mathematics and second of cosmology. [5]

The ancient Greek writers, who first developed both philosophy and mathematics, defined the concept of infinity as apeiron that means, literally, unbounded. In so doing, they gave what they thought was an unequivocal method for distinguishing the infinite from the finite: Something is infinite if it has no limit or end; it is boundless, or unlimited, or endless; it is chaotic and lacks any structure or order. In essence, they defined infinity by contrast to the finite: The infinite is totally different from, even opposed to, those things that make up ordinary experience.

We might balk at this definition, since there are many things in the world and in mathematics that have no end or limit or boundary but that are still finite. Think of the surface of the earth or the circumference of a ring: Each has a definite size but no "edges" or intrinsic limits. Or think of the length of a line segment to see the irony here: the segment defined by $0 \leq x \leq 1$ is finite since it is bounded at both ends (0 and 1), whereas the smaller interval $0 < x < 1$ is unbounded. But surely the infinite is not less than the finite! Or is it?

Actually, paradoxes related to infinity such as these date back as far as ancient Greece, traceable to Zeno of Elea (490-430 B.C.E.). The classical example is the famous race between Achilles and the tortoise. Being slower, the tortoise is given a head start over Achilles; but being faster, Achilles overtakes the tortoise and wins the race. But how is this possible, since every time Achilles catches up to where the tortoise was, the tortoise has moved ahead a small amount, and so on, and so on. . . . How is it possible for Achilles to pass the tortoise in a finite amount of time?

Through arguments like these, classical Greek thought came to view the finite world of our experience in sharp contrast with the infinite. Thus, they believed that a formless and indeterminate infinity underlies the world of finite entities and that finite entities are good precisely because they are formed, bounded, and determinate in contrast with the infinite.

Speculation concerning infinity traces back even farther in time. Anaximander (610-547 B.C.E.) was probably the first Western philosopher to think in detail about infinity. The infinite was a spherical, limitless substance: eternal, inexhaustible, lacking boundaries or distinctions. The world of finite entities arises out of that which is the infinite. Pythagoras (570-500 B.C.E.), too, rejected the infinite as having anything to do with the real world. All actual things are finite and representable by the natural numbers 1, 2, 3, and so on. Pythagoras taught that the geometrical forms—point, line, plane, and solid—arise when a mathematical limit is imposed on the underlying infinite structure. Plato (428-348 B.C.E.) believed that the Good must be definite rather than indefinite, and therefore it must be finite rather than infinite. According to Plato, God as the demiurge imposes limitations (i.e., intelligible form) on pre-existent matter, giving rise to the structured world around us as an ordered whole instead of a formless, unintelligible (ie., infinite) chaos. In all these cases, the infinite took on a negative quality compared with the finite.

But it was the philosopher Aristotle (384-322 B.C.E.) who provided a vital part of the conception of infinity that continued to pervade Western thought until the nineteenth century. [6] Aristotle altered the concept of infinity to mean an unending process, something unfinished or incomplete: ". . .(T)he infinite has this mode of existence: one thing is always being after another, and each thing that is taken is always finite, but always different." [7] For example, consider the unending succession of natural numbers 1, 2, 3, . . ., the endless succession of the seasons, or the endless division of a line interval into smaller and smaller portions. [8] Aristotle apparently thought of these as potentially infinite, since the series can be continued endlessly, but not as fully or actually infinite, since being endless we never really transcend the finite steps in the series to reach the term or limit of the series, i.e., the actually infinite, and thus to view the infinite series as a given totality. "It is plain from these arguments that there is no body which is actually infinite." [9]

This view in turn was altered further by the philosopher Plotinus (205-270 A.C.E.), according to whom God, or the One, is infinite, being free of both internal and external limits.[10] Plotinus saw matter as evil because it is infinite, unbounded, and formless, following the earlier philosophical tradition of the Greeks. But he also thought of the divine mind as good because of its inexhaustible power, overwhelming unity, and self-sufficiency. Still, the absolute could not be considered infinite given the pejorative connotations of *apeiron*. Instead, the absolute is beyond conception or analogy with the finite world. Here, then, for the first time the infinite took on a positive quality, although it still carried the connotation of the unbounded and unlimited. With Plotinus, the notion of the infinite moves between that of Plato and that of the Christian tradition, for whom the first attribute of God is infinity.

Early Christian writers were informed by this positive conception of infinity as they developed a doctrine of God based on Hebrew and Christian scripture, and in time the attribution of infinity to God became a standard cornerstone of Christian theology. Augustine (354-430 A.C.E.), for example, was heavily influenced by Plotinus. He believed that since God is infinite, God must know all numbers and must have limitless knowledge of the world.[11] Still, the category of the infinite served to radically distinguish God from the world. Writing a millennium after Augustine and drawing on the philosophy of Aristotle, Thomas Aquinas (1225-1274 A.C.E.) believed that only God is infinite. All creatures are finite, and even God, whose power is unlimited, ". . . cannot make an absolutely unlimited thing. . ."[12]

The infinity of God also has been widely assumed by philosophers and theologians in modern periods, although the meaning of God has differed widely. For example, according to such pantheists as Giordano Bruno, Baruch Spinoza or G.W.F. Hegel, God and the world are both infinite and, ultimately, both identical. On the other hand, in both classical and contemporary theism, a fundamental distinction is made between God and the world. Theists affirm that although God creates and sustains the world in existence, God utterly transcends the world. Thus, God alone is infinite, and the world is finite. God's infinity is a mode of God's perfection and God's goodness, and at the same time, God is incomprehensible.

We can know *that* God is perfect, infinite self-existence, but we cannot conceive of *how* God is perfect, infinite self-existence. Contemporary "pan*en*theists" tend to agree with theists about the infinity of God, but stress the immanence of God as well: The world subsists in and is sustained by the power and reality of God even while God radically transcends the world as its source and destiny. Examples include Arthur Peacocke, Ian Barbour, Jurgen Moltmann, as well as process theologians.

The central point here, however, is that the concept of infinity was transformed from a negative to a positive quality by early Christian theology and almost universally attributed to God as the supreme source of being, truth, and goodness. Still, infinity was defined in terms of a contrast with the finite or in terms of a goal that the finite never attains.

Moving from the context of philosophy and mathematics into cosmology, we find that the early Greek universe was seen as a mixture of finitude and infinity. According to Plato, the Earth was the center of the universe. The stars were fixed on a sphere that rotated about the Earth daily. What about the motion of the Sun, the Moon and the planets relative to the fixed stars? Greek astronomers attempted to solve this problem by adding concentric spheres with axes angled to compensate for the planetary motions. By the time Aristotle proposed his model of the heavens in *De Caelo* (On the Heavens), fifty-five transparent spheres were needed to accommodate the flood of increasingly accurate observations. Still, the universe was finite in size, although Aristotle believed it had existed forever and was thus infinite in the past.

Four hundred years later, Claudius Ptolemy (circa second century A.C.E.) was to incorporate a series of epicycles, deferents, and eccentrics to give even higher predictive accuracy. This system was, it seemed, endlessly adjustable, lasting until the great revolution of the sixteenth century with the new heliocentric cosmology of Nicolas Copernicus (1473-1543). Moreover, Ptolemy's model was essentially a geometrical construct, a mathematical device highly accurate but unrealizable in nature, whereas Aristotle's model was thought to represent the physical structure of the heavens.

What is important for our purposes here, of course, is that all these developments from the ancient Greeks to the Copernican revolution saw the universe as finite in size. They differed over the question of the age of the universe. Although Aristotle believed the universe to be eternal, Christian theology claimed it to be created by God and thus finite in age. Augustine was the clearest on this of the Patristic theologians. It is not that God created the world at some moment in a pre-existing time. Rather, God created time as well as matter.[13] By the thirteenth century, Aquinas was to resolve the debate by arguing that the eternity of the universe cannot be settled by philosophy (i.e., by appealing to Aristotle). Instead, its finitude in time is given us by revelation (e.g., Genesis 1). Thus, the conception of the universe at the time of the Copernican revolution was one that is finite in size and age.

Infinity in the Rise of Science: Examples from the Mathematics of Galileo and the Cosmology of Newton

The beginnings of the modern understanding of the mathematical concept of infinity can be traced to Galileo Galilei (1564-1642). In a delightful section of *Two New Sciences*,[14] Galileo argued that, as far as the real numbers are concerned, infinite sets do not obey the same rules as do finite sets. Consider the unending sequence of whole numbers: 1, 2, 3, . . . and the unending sequence of the squares of the whole numbers: 1, 4, 9, ... Intuitively we think there must be more whole numbers than squares since the set of whole numbers contains the squares along with other numbers (e.g., 2, 3, 5, 6, 7, 8, 10, . . .). Yet the squares can be put into a one-to-one correspondence with the whole numbers:

$$\text{whole numbers:} \quad 1 \quad 2 \quad 3 \ldots n \ldots$$
$$\downarrow \quad \downarrow \quad \downarrow$$
$$\text{squares:} \quad 1 \quad 4 \quad 9 \ldots n^2 \ldots$$

Now, suppose we define two sets of numbers as equal if and only if each number in set A can be associated with a number in set B, and nothing is left unmatched in sets A or B. Adopting this definition we find that the series of squares 1, 4, 9, . . . is equivalent to the series of whole numbers 1, 2, 3, . . . ! This seems to mean that there are as many squares as there are whole numbers! Given this paradoxical result, Galileo recognized that unlike finite quantities, "the attributes 'larger,' 'smaller,' and 'equal' have no place . . . in comparing infinite quantities with each other. . ."[15] We shall see that Galileo's insight was to play a key role in the discovery of "transfinite" numbers by Georg Cantor (1845-1918) in the nineteenth century.

Although Copernicus gave us a heliocentric framework, it was Isaac Newton (1642-1727) who first saw the possibility of a scientific cosmology in which the universe is infinite in size and eternal in age. Newton's cosmology was based on his theory of mechanics, his (metaphysical) assumptions about absolute space and time, and his theory of gravity. Newtonian mechanics interpreted changing motion in nature as the result of impact forces between masses: An acceleration, a, of a body, B, is produced by a force, f, inversely proportionate to the inertial mass, m, of B, or f=ma. For Newton's system to work, however, one must distinguish between real and apparent acceleration, and to do this Newton invoked what he called absolute space and absolute time. Both are infinite and passive, giving to masses their relative acceleration without being altered by the changing states of matter in the world. Newton also proposed a theory of gravity in which any two masses in the universe interact with each other with a force proportional to the product of their masses and inversely proportional to the square of the distance between them.

Turning to cosmology, where only gravity is involved, Newton faced a quandary: If the number of stars in the universe is finite, why don't the stars collapse into a single mass under their mutual gravitational force? If the number of stars is infinite, allowing them, possibly, to remain distributed evenly (pulled evenly in all directions), why isn't the night sky ablaze with starlight in every direction (i.e., Olber's paradox)? For these and other problems, Newton's cosmology remained an ambiguous influence on culture. It was widely adopted as a description of the universe as infinite, eternal, and unchanging, yet its technical scientific problems were left unsolved until the work of Albert Einstein in the early twentieth century which transformed physics and cosmology.

The Discovery of the Transfinite in Modern Mathematics

Galileo's insights into infinity remained dormant until the discoveries of Bernard Bolzano (1781-1848), Richard Dedekind (1831-1916), and Cantor. Before their work, the infinite was thought of as merely a potential infinity, the limit of an endless series, something never fully achieved either in mathematics or in nature. The birth of the calculus in the seventeenth century involved complex questions about the status of infinitesimals both as minute extensions in space and durations in time. Yet even Newton, Gottfried Wilhelm Leibniz (1646-1716) and Carl Friedrich Gauss (1777-1855) adhered to the view that such infinities and infinitesimals are only abstractions or potential infinities.

At the close of the nineteenth century, however, Cantor discovered that there are many kinds of mathematical infinity, and he explored their structures.[16] To do so, Cantor first showed how to apply basic mathematical principles, which we use with finite sets, to infinite sets and principles including addition, multiplication, exponentiation, and the relation "greater than." This in turn meant he could give an explicit procedure for constructing different kinds of infinity and for showing their internal consistency. In essence, we could say that infinity, instead of being defined simply by contrast with the finite and remaining essentially unknown, has now been explored and we know a great deal about it—even an infinite amount!

To see this, we have to be clear about some basic ideas we encounter on a daily basis. Let's start with counting and from there see what we mean by saying that two (finite) sets are equivalent.

When we count something, we do so by placing each element in the set of things to be counted in a "one-to-one correspondence" with the natural numbers $1, 2, 3, \ldots$. For example, suppose I want to count the number of marbles on the table in front of me. I count them one at a time, saying "one" as I touch the first marble, "two" as I touch the second, "three" for the third, and so on until I come to the last marble. The number I associate with the last marble as I touch it, say "nine," is the number I assign to the handful of marbles I began with. In counting the marbles and assigning them the number "nine," I implicitly think of the marbles as a complete set and I associate the number nine with that set. Mathematicians call the number nine the "cardinal number" of that set.

Similarly, I could count the number of coins in my pocket. If I found I had nine coins, then I can think of the set of marbles and the set of coins as equivalent: Since I can put the elements in each set in a one-to-one correspondence with the elements in the set of natural numbers and obtain the cardinal number nine in each case, then I can put the set of coins and the set of marbles into a one-to-one correspondence with each other. Indeed, all sets of objects of any kind that carry the cardinal number nine are equivalent. Moreover, the cardinal number for all these equivalent sets distinguishes them from all sets with the cardinal number 8, 34, 5002, and so on.

Now we make Cantor's fundamental claim: The concepts of "equivalent sets," or "counting," and of "cardinal number" can be transferred from their foundations in finite sets to infinite sets. Cantor coined the term "transfinite" for the cardinal number of infinite sets, for reasons that become clear shortly. Rather than assuming the infinite is in direct contrast with the finite, as had always been done in the past, Cantor treated an infinite set in a direct analogy with how he treated finite sets. The results are staggering!

To see this, we start with the set of natural numbers $\{1, 2, 3, ...\}$, which serves as our simplest example of an infinite set. Cantor chose $\aleph_0$ (pronounced aleph-null) to represent its transfinite cardinal number. He thought of $\aleph_0$ as a whole, and not just an incomplete sequence of unending numbers ascending in scale. In other words, Cantor actually distinguished between an unending sequence of elements, such as the sequence $1, 2, 3, . . .$, which is potentially infinite but always, in fact, finite, and the complete infinite sequence thought of as a whole, i.e., the set $\{1, 2, 3, . . . \}$. He called the potential infinite a "variable finite" and symbolized it as ∞; the actual infinite he symbolized by $\aleph_0$, as we saw above. Thus, ∞ never reaches completion, never becomes $\aleph_0$. To think about ∞ is to think of an ever-increasing series of numbers continuing forever without reaching an end. To think about $\aleph_0$ is to stand outside this series, from God's point of view as it were, and to consider them as a single, unified, and determinate totality.

He then extended what we know about counting finite sets to infinite sets: all infinite sets whose elements can be put in a one-to-one correspondence with the natural numbers will have the same cardinal number, $\aleph_0$. We call these sets denumerably infinite or countably infinite. This leads to some really surprising results. For example, recall Galileo's paradox about square numbers. Since there is a one-to-one correspondence between the natural numbers and the square numbers, it means that the set of square numbers is uncountably infinite; it has the same cardinal number, $\aleph_0$, as the set of natural numbers. Similarly, the set of even numbers is equivalent to the set of natural numbers, as is the set of odd numbers.

What is even more surprising is that, according to Cantor, we can generate infinities that are "bigger" than the set of natural numbers in a certain sense, although all these still have cardinal number $\aleph_0$. Here it will be helpful to introduce the term ordinal number to represent the order of the elements in a set. For example, the finite set $(1, 2, 3)$ has cardinal number 3, since there are three elements in the set, and it has ordinal number 3, since the set, as an ordered group of elements, has a third element. In the case of finite sets, the ordinal number and the cardinal number of a set are the same.

It turns out, though, that they are not the same for infinite sets! We start, as above, with the infinite set of natural numbers, $\{1, 2, 3, \ldots\}$; following mathematical custom, we designate its ordinal number as ω. Now let us think of this set as a whole, as complete in itself. If we do so, then it is possible to conceive of adding 1 to the set, forming a new set, $\{1, 2, 3, \ldots, 1\}$; the ordinal number of this set would be $\omega + 1$.* Adding 1 to this set yields another new set, $\{1, 2, 3, \ldots, 1, 2\}$, whose ordinal number is $\omega + 2$.

As we continue the process, we generate a whole ladder of increasingly complex infinities. For example, if we add the set of natural numbers to the set of natural numbers, we obtain the set $\{1, 2, 3, \ldots, 1, 2, 3, \ldots\}$ whose ordinal number is $\omega + \omega$, which we notate as $\omega.2$. Continuing from here, we can consider $(\omega.2) + 1$, $(\omega.2) + 2$, $(\omega.2) + 3$, and so on until we reach $\omega.\omega$ which we can write as ω^2. Again we continue adding to this set until we conceive of ω^3, ω^4, and so on. This in turn points toward its goal: ω^ω—but there's still more, in fact infinitely more! We can think of an infinite series of exponential powers, raising ω to the ω power infinitely many times. What is even more astonishing is that the elements in any of these transfinite sets can be put in a one-to-one correspondence with the elements in the set of natural numbers, $\{1, 2, 3, \ldots\}$! This means that all these sets, even though differing in their ordinal number, are denumerably infinite:

---

*Note that $\omega + 1$ is not equal to $1 + \omega$. The latter is still equal to ω. That is, $1 + \omega = \omega$ since all the symbol $1 + \omega$ means is that we add one element to the elements that taken endlessly but thought of as a whole is the set of natural numbers whose ordinal number is ω. The former, $\omega + 1$, means adding to a given infinite whole, namely $\{1, 2, 3, \ldots\}$, a new element 1, thus forming the set $\{1, 2, 3, \ldots, 1\}$. This means that $\{1, 2, 3, \ldots, 1\}$, taken as a whole, is not equivalent to the set $\{1, 2, 3, \ldots\}$, taken as a whole. Thus, $\omega + 1 \neq \omega$ although $1 + \omega = \omega$.

they have the same cardinal number, $\aleph$! Mathematicians express this fact by noting that

$$\aleph_0 + \aleph_0 = \aleph_0$$

even though it remains true that

$$\omega + \omega \neq \omega$$

Apparently, the rules that infinity obeys are both like and unlike the rules that finite sets obey!

But there are still more surprises ahead, since we can imagine sets whose "infinity" is so great, as it were, that they cannot be put into a one-to-one correspondence with these sets: they are "uncountably infinite." In 1874, Cantor proved this for the set of real numbers. (Real numbers are composed of the natural numbers, the fractions, and the irrational numbers.) Since the real numbers can be put in one-to-one correspondence with the points of a straight line, Cantor called the cardinal number of the set of real numbers "the power of the continuum," often designated by the letter c. It is surprising, although it is easy to show, that the number of points in any line of any length is the same. What is really surprising is that there is a one-to-one correspondence between the set of points in a plane and the set of points in a line, since one might well have thought the former to be infinitely greater than the latter. Cantor then extended this result to the points in a three-dimensional space, and then to a space of any number of dimension. All these mathematical objects—the line, the plane, the volume, the volume in four or more dimensions—have the same cardinal number, c.*

---

*We can go further still, although a proof lies beyond the limits of this chapter. Cantor showed that one can construct a whole series of transfinite cardinal numbers $\aleph_0$, $\aleph_1$, $\aleph_2$, and so on, leading to $\aleph_\omega$. Even this is not the end, though. We can think of $\aleph_{\omega+1}$, $\aleph_{\omega+2}$, . . ., $\aleph_\omega\omega$, $\aleph_{\aleph_0}$ and so on—there's never an end to the kinds of infinity we can construct. At the same time, these infinities share an important feature with finite sets since, no matter how complex they seem, the transfinite forms of infinity are at least conceivable by construction. We are thus in a better position to understand why Cantor called them "transfinite."

We are also prepared—somewhat—to take the final step and consider what lies beyond even the transfinite numbers: Cantor called it "Absolute Infinity" symbolized as Ω . In one sense the Absolute Infinity is inconceivable; it is beyond rational understanding. Yet in another sense we can know something about Ω : It exists as a coherent mathematical concept. We can even know something about its properties! To make this apparent contradiction make sense, consider the converse: [17] Suppose Ω is conceivable. Then there must be some property P that is exclusively a property of Ω and Ω alone, so that we can conceive of Ω as the only number that has property P. Now, to make Ω inconceivable, we merely have to stipulate that every property P is shared by both Ω and some transfinite ordinal. If so, then there is no property P that is unique to Ω and not shared by Ω and some transfinite ordinal. Thus we can know the properties P of Ω, since every conceivable property P is shared by some transfinite ordinal. Yet because of this, we can never differentiate Ω completely from the transfinite ordinals, since we can never describe Ω as possessing a property P that it doesn't share with a transfinite ordinal. In essence we can never tell Ω from some transfinite ordinal. We can never know if we are conceiving of Ω and not some transfinite ordinal. In short, we can never conceive of Ω alone. Therefore Ω is in itself inconceivable. Ω is inconceivable because Ω can never be uniquely characterized or completely distinguished from something that, although infinite too, is a lower order infinity than Absolute Infinity. The transfinite numbers, Cantor's endless types of infinities, lead towards Absolute Infinity Ω but never reach it, since Ω lies beyond all comprehension. This argument is often called the Reflection Principle.[18]

In the end, Cantor set up a threefold distinction regarding the infinite: the Absolute Infinity that is realized in God and the transfinite numbers that occur both in mathematics and in nature. Rucker quotes Cantor as follows:

> The actual infinite arises in three contexts: *first* when it is realized in the most complete form, in a fully independent other-worldly being, *in Deo,* where I call it the Absolute Infinite or

simply Absolute; *second* when it occurs in the contingent, created world; *third* when the mind grasps it in *abstract* as a mathematical magnitude, number, or order type. I wish to mark a sharp contrast between the Absolute and what I call the Transfinite, that is, the actual infinities of the last two sorts, which are clearly limited, subject to further increase, and thus related to the finite.[19]

We will pursue the question of the relation between the absolute in mathematics and God below. But first, what about Cantor's claim about actual infinities in mathematics and in the world? These questions lead the theme of this chapter, since if an actual infinity occurs in either mathematics or in nature, it would seem to challenge the distinction between the finite and the infinite that has traditionally served as one of the leading ways theologically and philosophically to describe the world as created and to describe God its creator, i.e., to distinguish God and Creation.

Does an Actual Infinity Make Sense in Mathematics or in Cosmology?

Is Cantor correct in believing that the world is actually infinite? With this, we open onto one of the central questions in this essay. First, do mathematicians in general agree with Cantor on the claim that infinity is a consistent and coherent concept in mathematics? Next, do cosmologists think that the universe is actually infinite in size and will it last forever?

The first question receives a mixed response. Many contemporary mathematicians accept Cantor's theories as valid, but they reject the applicability of transfinite numbers to realms beyond mathematics. In a recent survey of the issue,[20] William Craig cites Bolzano, who claimed that infinite sets, such as the set of all "absolute propositions and truths," are infinite, but sets such as these exist in the realm of mathematics alone, or what Bolzano called the "realm of things which do not claim actuality, and do not even claim the possibility" of actuality. According to Craig, Bolzano gives God, as his only example of a real actual infinity.[21] Craig gathers additional support for this view from Abraham Robinson, Abraham Fraenkel, Alexander Abian, Pamela Huby, and David Hilbert.[22]

Taking the problem one step further, Craig also describes the problems that beset Cantor's theory in the decades that followed his initial discoveries. These problems undercut a "naive" view of the existence of actual infinities even in the realm of mathematics.[23] In Craig's view, "even the mathematical existence of the actual infinite has not gone unchallenged and therefore cannot be taken for granted...Therefore we conclude that an actual infinite cannot exist."[24]

When we turn to physics and cosmology, however, a different answer lies in store. We should first note in passing that contemporary physics is filled with mathematical infinities. Even classical fields go infinite at their point sources, but quantum fields entail a staggering variety of infinities: the vacuum energy is infinite, charge and mass in quantum field theory require renormalization to make them consistent with measurement, the production of matter/antimatter pairs and their mutual annihilations and the numbers and kinds of particles exchanged during interactions are endless, and so on. Perhaps these are all strictly mathematical artifices, symptoms of the transitory nature of physical theory or problems that will be removed by improvements in future theories. Perhaps not. The question of the status of infinity in physics is immense.

But here we return specifically to the theme of cosmology, where according to Einstein's general theory of relativity, space and time curve like an elastic continuum distended by matter as it moves about the universe. At the same time, the trajectories of moving matter are shaped by the curvature of the space-time in which matter moves. The story is most striking, however, when we consider the universe itself. According to big bang cosmology, the universe is expanding from an origin some 15 or so billion years ago described colloquially as the beginning of time, or "t=0." This initial event is a genuine "singularity," a space-time point characterized by infinite density, infinite temperature, and zero size. But is this really a description of nature as it was at the "first moment," t=0, or is it merely a quirk of the mathematics that future theories will overcome?

Big bang cosmology also portrays the universe as either closed, shaped like a three-dimensional sphere with a finite size that changes in time (first expanding then contracting), or open, shaped like a three-dimensional saddle with an infinite size that expands forever. What about these features of the universe: Are they real, so that the universe may be actually infinite? Or are they, too, the spurious product of mathematics and irrelevant to nature as it actually is?

Clearly, the initial singularity, t=0, may be replaced eventually by a new cosmology in which the universe has an infinite past. For example, in some versions of inflation, the problem of t=0 is overcome, along with other technical problems troubling the original big bang account. More recently, some proposals in what is called "quantum gravity" move in this direction, as popularized in Stephen Hawking's book, *A Brief History of Time*.[25] Whether this undercuts the question of the "finite age" of the universe theologically is a matter of some debate.[26] More interesting for our purposes is the question whether the universe is open or closed. (Both possibilities still hold even in the case of quantum cosmology: After the initial moments of the universe, the quantum picture must lead to the standard big bang scenarios.) Most astronomical evidence to date suggests that the universe is open. What then are we to make of the claim that the universe, being open, is spatially infinite?

Actually this claim raises even more issues. For one thing, such a universe, although spatially infinite, is still expanding in time! Markers floating in space move apart from one another, and there is an actual infinity of markers in an endless spatial expanse. For another, the geometry of the open model, unlike that of the closed, spherical model, cannot be embedded in Euclidean space. What this means is that the closed model can be pictured as a three-dimensional sphere without much loss of accuracy, but the open model cannot really be pictured as a three-dimensional saddle. At each point in space, the curvature is like that of a saddle, but the way space curves at each point extrudes out away from space-time; it is something like a three-dimensional mountain that stretches out away from a two-dimensional topological map, although the map's contour lines attempt to suggest the distortion. In this sense, language about an open universe is apophatic; it is kataphatic about a closed universe. Who would have guessed?

What does this model say about the status of infinity and its bearing on our understanding of the infinity of God? We turn to this question, along with those drawn from Cantor's astonishing work on the mathematics of infinity, in the final section below.

Enriching our Understanding of the Infinity of God

We've covered a lot of territory, and yet we've barely scratched the surface. A tremendous amount of further research awaits—perhaps tapering off to infinity! Yet even with this initial foray, we can begin to collect some gems found in the process that seem worthy of further reflection. To do so, we need to recall our original discussion about the meaning of infinity. There I suggested that the early Christian world helped to transform the meaning of infinity from its negative connotations involving what is an unlimited chaos, a gnawing privation, to more positive connotations suggesting ultimate reality, the ground of being, the highest good and the source of the world. At the same time, however, theologians retained the classical Greek distinction between the infinite as wholly different from the finite and in contrast to it.

Thus to say that God is infinite is really to say that we cannot comprehend God. This distinction, inherited from the Greek philosophical culture by the early church, has predominated through the centuries in Christian theology as it seeks to speak about God. We see this most clearly in the distinction theologians make, and which we discussed at the onset, namely between the apophatic and the kataphatic. We start with acknowledging the apophatic, i.e., how little we know of the unseen, incomprehensible mystery that is God. Then we move to the kataphatic, i.e., we seek to express something that we do know about God: that God intends to be known and that what we know is God's love, care, purpose and mercy, and so on. Moreover, the term "infinity" has played a pivotal role in this history: It has been used almost exclusively to express the stark difference between the apophatic and the kataphatic. God is infinite; we are finite. God is other than, unlike, wholly different from us. When we claim that God is infinite, we intend to mean something apophatic: God is beyond, inscrutable, totally other than the known, the finite world as we know it.

Thus, if any property of what we think of as the finite world would turn out to be, in reality, infinite, it would seem to challenge our understanding of God: Either the world, if infinite, would be like God, thus leading to pantheism, or the world, if infinite, would simply be the infinite world and its own best explanation and there would be no need for God, thus leading to atheism. The importance of the finitude of the world has resided in the fact that it has been a key defense against unbelief as well as a key constituent to the Christian distinction between Creator and creation.

In short, this means that an infinite universe, either existing forever from eternity or having an infinite size, would seem to challenge Christian faith. Given that one model in the big bang theory is an open universe of infinite size and that will continue to exist for eternity, it would seem that big bang theory would challenge Christian faith.

One might reply, of course, that if the universe is actually infinite, this simply means we need to adjust our concept of God, that our traditional idea of God is too small and that God is, in fact, "more infinite" than even an infinite universe. This may be the case and here Cantor's insights about the Absolute Infinite versus the merely transfinite could be very helpful.

Alternatively, one might reply that the meaning of infinity in theology is highly ambiguous or purely metaphysical; therefore it cannot directly challenge the meaning of infinity in cosmology. Although he sees the implications of mathematical infinity for theology, Timothy Pennings highlights the ambiguity of the term nicely: "To say that

God is infinite could be referring to some attribute of God such as His omnipresence (Locke), or it could be making an absolute statement about God's being (Aquinas), or it could be contrasting God's nature with human nature (Weyl), or it could be linking God's knowledge to the set of (natural) numbers (Augustine), or it could be associating the mind of God with the Absolute Infinite (Cantor)."[27] But my argument has been that despite these ambiguities, there is a historical continuity in the way infinity has been used by scientists, philosophers, and theologians and a common thread in its many meanings. It is this common thread that makes the challenge from cosmology serious.

Now we are in a position to appreciate the radical importance of Cantor's discovery of the complexities of infinities in mathematics and to apply them, at least by analogy, to our problem. Clearly, as we pursue these leads in the future, we will have to take into account the serious problems arising from the antinomies in set theory (see Note 23). I want to emphasize this very clearly, but I also want to stress that these problems do not rule out using Cantor's work by analogy in the theological domain of discourse. What I will suggest is that Cantor's work, at least by analogy as we move across fields, enhances the meaning of God as Creator and gives new insight and energy to the proclamation of faith in God the Creator. To see this, let us turn first to mathematics and then to cosmology.

Mathematics

The Reflection Principle in mathematics points to a fascinating connection between what we know and what we do not know about Cantor's Absolute Infinity. In theological language, we would say that the Reflection Principle intertwines what was otherwise kept distinct, the apophatic and the kataphatic, or the way of negation and the way of affirmation. This alone is alluring.

We begin with Cantor by defining the mathematically Absolute Infinity as beyond comprehension, lying beyond the unending ladder of the transfinite numbers that are themselves infinities but that can be fully comprehended, ie., whose properties can be formally described. The way we ensure that the Absolutely Infinite is beyond comprehension is by claiming that if it were not so, we could describe it unambiguously, and by describing it, comprehend it. Now to describe it means to state at least one property that it alone possesses and that allows us to single it out from the transfinite numbers. We therefore reverse this and insist, via the Reflection Principle, that all its properties must be shared by the transfinites. This ensures that the Absolute Infinity cannot be uniquely described. Since none of its properties is unique to itself, we can never point unequivocally to it by pointing to that unique property. In this sense, Absolute Infinity is indescribable and incomprehensible.

But this, in turn, means that we do in fact know something about the Absolutely Infinite: All the properties it possesses must be shared with and disclosed to us through the properties of the transfinites. The Absolute Infinite is in this sense knowable, comprehensible; each of its properties must be found in at least one transfinite number. The Absolute is disclosed through the relative, or transfinite, infinities, and yet it is through this disclosure that it remains hidden, ineffable, incomprehensible.

Another way of putting this is that the incomprehensibility of Absolute Infinity is manifested by its partial comprehensibility. What we know about the Absolute Infinite is never more than partial knowledge shared by all relative infinities. What is truly unique about Absolute Infinity is never disclosed, but forever hidden. What we do know about Absolute Infinity forms a veil hiding what is forever beyond our knowledge. Yet what the veil that hides Absolute Infinity discloses something about Absolute Infinity. We endlessly learn more and more through endless circles of discovery of the transfinite, endlessly moving to more and more knowledge of that which can never be known exhaustively.

This situation of the mutual sharing and the intertwining of comprehensibility and incomprehensibility seems to apply preeminently well, even if only by analogy, to our knowledge of God. On the one hand, we want to affirm that God is both absolute mystery, the ineffable source lying beyond comprehension, and yet we want to affirm that we can know this God as Creator and redeemer. How can this be? It turns out that the theological response is almost an exact parallel with the mathematical response. The incomprehensibility of God includes the infinity of God as the apophatic tradition suggests. Yet, as the kataphatic tradition suggests, this God intends to be known by us and to know us. This God who would be known is known to us as our Creator precisely through God's self-revelation in Scripture and prayer and through the product of God's creative activity, the universe.

Now comes the analogy between the way mathematicians understand infinity and theologians speak about God: I propose that God, as Absolute Infinity, is hidden precisely by the sharing of all God's attributes with us through all that God creates. The Absolutely Infinite in mathematics is hidden precisely by the fact that it shares its properties with all the transfinites. The incomprehensibility of the Absolutely Infinite in mathematics is safeguarded by its reflection in all the known infinities. Thus, theologically, the mystery of God is vouchsafed by its revelation in our experience of God and in the properties of all that God creates and science discovers. Or once again, the Absolutely Infinite is known through the known infinities and yet being so, remains unknown in itself. Theologically, the God who is known is the God who is unknowable. What God has chosen to disclose to us, the kataphatic—God's existence as Creator, God's goodness, love, truth, and beauty—is a veil behind and beyond which the reality of God is endlessly hidden yet endlessly revealed.

The analogy with mathematics also has immediate connections with cosmology.

Cosmology

According to contemporary cosmology, if the universe is open, it is actually infinite in size and continues to expand forever. Even the suggestion that the universe may be infinite in size challenges any naive view of God as somehow a God limited to terrestrial concerns or a God who is really just a superbeing and not the ground, source, and destiny of all that exists. The God of the big bang must be thought of at least in terms of infinite presence, infinite power, and infinite compassion, if this God is to be God of a universe of which even the visible portion is billions of light-years in size.

But does this undermine the distinction between God and the universe? Does it even go so far as to undercut the argument for God's existence? Although pantheists or atheists might believe so, I believe instead that this enhances our understanding of the meaning of God for two reasons.

First, science presupposes that the universe is contingent. The universe need not have come into existence; its sheer existence points beyond itself and is unanswerable by science. In this sense, the existence of the universe poses a question that lies beyond scientific explanation. Science leads in the direction of theism and away from pantheism or atheism, since the contingency of the universe is precisely what theology assumes by calling the universe "creation." This, in turn, leads to the concept of God the Creator.

The second reason starts with what we have just said about the Reflection Principle in mathematics. This time we apply the Reflection Principle to the context of scientific cosmology. Does an infinite, open universe disclose something about the meaning of the infinite nature of God? Mathematically, the infinity of the universe must be transfinite and not absolute, for it can be specified by mathematical formalism (such as Einstein's equations). But now we apply the Reflection Principle in reverse, moving from the known infinities to the absolutely infinite that is beyond knowledge. That is, according to the Reflection Principle, the mathematical properties of the transfinite universe must be shared by God as the absolute infinite, yet in such a way that the universe as transfinite is never confused with God in God's Absolute Infinity. Thus, the infinity of the universe in terms of space and time reveals something of the God who is its source while at the same time they hide God, leaving God as unknown and incomprehensible.

Thus, in both cosmology and in mathematics, the discoveries of science about the subtle nature of infinity lead to a profound understanding of the belief that God is known through nature as her Creator precisely as that God is veiled by nature. Surely Anselm of Canterbury (1033-1109) pointed in the right direction, "God is that than which nothing greater can be thought." [28]

Conclusion

It is fitting that I close this essay by citing passages in which John Marks Templeton talks about God quite similarly to my general direction:

> (T)he humble approach . . . takes the position that God is infinite, whereas created objects are contingent and finite. Even the vast universe falls infinitely short of what God is . . . the Theology of Humility . . . proposes that the infinite God may not even be describable adequately in human words and concepts and may not be limited by human rationality . . . [29]

Cantor's work, as we have seen, takes up the problem of what we can say, and what we cannot, using human rationality about infinity. Moving beyond the previous centuries of reflection on infinity, Cantor has shown us that absolute infinity in mathematics utterly transcends even the infinite ladder of the transfinite infinities that he discovered and that may be embodied in the universe. Yet absolute infinity is revealed and known through the transfinite. It would seem then that the discoveries of science and mathematics about the universe point beyond themselves to horizons and realities whose ultimate source is the incomprehensible God. Will we follow these pointers, although keeping sharply aware of their profound limitations? Or will we be dazzled by science and stop short of that great, further exploration?

I am reminded once again of Newton with his seashells, holding the hard won gains of his science up to scrutiny, yet barely aware of the endless and unfathomable mystery beyond. I am also reminded of the humility of those early theologians who knew that when we seek to speak of God, we do so only out of the glimmers of understanding that sparkle amid the vast background of uncomprehended mystery, a mystery that nevertheless shines in nature and in the human spirit with unquenchable light.

Notes

1. Sir David Brewster, *Memoirs of the Life, Writings, and Discourses of Sir Isaac Newton*, Vol. ii (Edinburgh: Thomas Constable; Little Brown, 1855) Chap. xxvii.
2. As Owen Thomas points out, "although infinity is not a biblical concept, its various meanings are implied about God throughout the Bible." It includes God being present everywhere and at all time and being unlimited in power and knowledge. He cites Deut. 33:27; Ps. 90:2, 4; Ps. 139:7-12; Ps. 147:5; Job 38-41; Isa. 40: 8; Isa. 51:6; Rom. 1:20; 1 Tim. 1:17; 1 Pet. 5:10. The traditional difficulty has been to reconcile the concept that God is infinite with the concept that God is personal. Owen C. Thomas, *Theological Questions: Analysis and Argument* (Wilton, Conn.: Morehouse-Barlow, 1983) p. 38.
3. For readable and insightful introductions I recommend Rudy Rucker, *Infinity and the Mind: The Science and Philosophy of the Infinite* (New York: Bantam, 1983), and William Lane Craig and Quentin Smith, *Theism, Atheism and Big Bang Cosmology* (Oxford, England: Clarendon Press, 1995). For a recent paper that explores themes in common with this essay, see Timothy J. Pennings, "Infinity and the Absolute: Insights in Our World, Our Faith and Ourselves" in *Christian Scholars' Review* (December 1993). In this fine paper, Pennings underscores the importance Georg Cantor felt concerning his work on infinity and the theological tradition. Pennings also stresses the ambiguity in the theological meaning of God's infinity and the qualitative difference between the infinite and the finite.
4. Rucker, *Infinity and the Mind*, p. 3.
5. These sections draw in part from Rucker and from Craig and Smith.
6. I am using *The Basic Works of Aristotle*, edited by Richard McKeon (New York: Random House, 1941). Aristotle began, as those before him, by considering infinity as a negative quality, i.e., as something lacking: ". . . (I)ts essence is privation . . ." See *Physics*, III.7.207b35.
7. Aristotle, *Physics*, III.6.206a26-30.
8. The only infinite considered possible is the potential infinite: something capable of being endlessly divided or added to, but never fully actualized as infinite. See Aristotle, *Physics*, III.4.204b1-206a8
9. Aristotle, *Physics*, III.5.206a5b. Italics are in the original text. It is helpful

here to distinguish between the following two ideas: For every number n, there is another number n+1; there is a number n beyond every number 1, 2, 3, . . . The first suggests that we can always find more numbers in the series 1, 2, 3 . . . and thus that the series is potentially infinite (or endlessly extendable). The second suggests that there is an actual number n beyond the series as a whole, suggesting that the series as a whole is actually infinite. This second concept of an actual infinity was foreign to the Greeks.

10. Plotinus, *Enneads,* V.5.11.

11. Augustine, *City of God* XII 18.

12. Thomas Aquinas, *Summa* Ia, 7,2-4.

13. In a now famous move, Augustine endorses the creation of time along with the world and rejects the creation of the world in time, in *Confessions* XI.13 and *City of God* XI.6.

14. Galileo Galilei, *Dialogues Concerning Two New Sciences,* trans. Henry Crew & Alfonso de Salvio (New York: Dover Publications, 1954). See "First Day," pp. 1-108.

15. Ibid., p. 33. We should note historically that underlying this paradox was the notion of space and time as continuously varying quantities and, consequently, the problem of finding the instantaneous velocity of a moving body. The latter involves the notion of an infinitesimal quantity of time, dt, as developed in the calculus of Newton and Leibniz. To avoid such problems, the notion of the actually infinite was replaced by the notion of a limit.

16. Craig gives a helpful presentation of Cantor's transfinite numbers in *The Kalm Cosmological Argument* (New York: Barnes & Noble, 1979) on which I will draw here. Also see "Infinity" by Hans Hahn in James R. Newman, ed., *The World of Mathematics* Vol. 3 (New York: Simon and Schuster, 1956), p. 1593-1611; Rucker, *Infinity and the Mind,* chaps. 1, 2.

17. I am following Rucker's explanation closely here: *Infinity and the Mind,* p. 53.

18. Rucker puts it this way: "Every conceivable property that is enjoyed by Ω is also enjoyed by some set . . . Every conceivable property of the Absolute is shared by some lesser entity." Ibid., p. 53

19. Georg Cantor, *Gesammelte Abhandlungen,* p. 378. From Rucker, p. 10.

20. Craig writes as follows: "(W)hile the actual infinite may be a fruitful and consistent concept in the mathematical realm, it cannot be translated from the mathematical world into the real world, for this would involve counter-intuitive absurdities." Ibid., p. 9. As is clear in what follows in my chapter, I do not agree with Craig here. Also see Craig, p. 9-11, for references to these works.

21. Craig is not entirely accurate here. Bolzano also allowed for what he called "a trace of infinitude" in the world around us. This includes the set of finite beings and the "conditions experienced by any single one of them during no

matter how short an interval of time . . . We therefore encounter infinities even in the realm of the actual." Ibid., p. 101.

22. Ibid., p. 12.

23. He first recounts four philosophical interpretations of the ontological status of mathematical sets: 1. Platonism (realism): Sets have a real existence; mathematics discovers, but does not create, the infinite. 2. Nominalism: Neither sets nor the numbers that are contained in them have any real existence in the world. 3. Conceptualism: Sets and numbers exist as mental objects, not as physical objects, created in the mind of the mathematician. 4. Formalism: Mathematics is merely a consistent way of calculating entities in the real world and nothing more. He then describes the problems that arose regarding Cantor's theory: Burali-Forti's antinomy, Cantor's antinomy, and Russell's antinomy. According to Craig, these antinomies forced major revisions in set theory (logicism, axiomatization, and intuitionism). Most important, they undercut the Platonist/realist interpretation, but not the others. Ibid., 16-24.

24. Ibid., p. 24.

25. See essays in Robert John Russell, Nancey Murphy, and C. J. Isham, eds., *Quantum Cosmology and the Laws of Nature: Scientific Perspectives on Divine Action* (Vatican City State: Vatican Observatory Publications and Berkeley: Center for Theology and the Natural Sciences: 1993).

26. Stephen W. Hawking, *A Brief History of Time* (Toronto: Bantam, 1988).

27. Pennings, op. cit.

28. St. Anselm's *Proslogion*, trans. M.J. Charlesworth (Notre Dame, Ind.: University of Notre Dame, 1979) p. 117.

29. John Marks Templeton, *The Humble Approach*, rev. ed. (New York: Continuum Publishing Group, 1995) p. 21, 165.

PART 3

Limits to Scientific and Theological Answers

The Voices of Theologians and Humanists

Martin E. Marty

Whoever hears How large is God? asked directly, or whose curiosity is piqued by those who ask it of others, will likely respond with other questions.

Why do you ask? is the natural first interrogative reply. Certainly, serious people who query others about the size of God do not mean to get opinions about the physical dimensions of a deity. Prophets in numerous religious traditions have often in the past derided those worshipers who reduced their god or gods to the size of portable household deities, usually made of wood or stone. Of course, outpost tribes or individuals—whether on primitive islands or in sophisticated metropolises—may even in the present worship idols or fetishes that are or can be reduced to shelf size. But such objects are not usually at issue in serious inquiries like this one.

If "large" here did refer to physical dimensions, one would simply approach venturesome or foolish physicists and mathematicians to speculate on the length of the tape measure it would take to encompass the billion billions of galaxies that may make up our universe. It would be absurd to think of them, or it would be to think of them as absurd, if such measurers then went on and merely added a fraction of an inch or more to their calculations. They would be figuring that God defined as Creator or prime mover or first cause would have to be physically larger than the universe in order to have created, moved, or caused it to be. Or the brain of the child in early stages of awareness may thus conceive God in terms of physical size. And some who deride those who conceive of God at all may also for the sake of argument or belittlement reduce the God-question to one of physical size. But such measurements also are not usually at issue in serious inquiries like the present one.

Passingly, but not so briefly, we should mention that the present context is further not one in which existential versions of the questions about the size of God are in place. From time to time, a religious thinker will write, as J. B. Phillips did some years ago, a book with a teasing title such as *Your God Is Too Small*.[1] Such works usually have to do with the limits some prayerful folk put on the power of a personal God, so that this God cannot be of much help to them, whether as judge or savior or comforter. While they may not use the precise formulaic version, How large is God? believers do ask something like this of others when some of them come to grief. Is the thou to whom you pray and whose comfort you seek great enough to hear and to provide support? How large is God? one may also ask in times of doubt or despair, when this thou seems remote or eclipsed, silent or dead. Is your God so small that you feel abandoned? Is your God so reduced that you can get no affirmations to counter your doubts or despair?

Such parts of interrogative conversations occur millions of times daily around the world, in backyards or sick rooms or counseling chambers, in the framework of any theistic religion. This version is posed to those who have special qualifications for speaking of God as the thou who would be known in existential contexts. Through the centuries these questioned ones have been pioneers and leaders who have claimed to hear the voice of the other. This would have been the case with Moses when, according to the Hebrew scriptures, he saw a burning bush that was not consumed and heard a voice. He had to know whether the voice belonged to the one who was great enough to lead if Moses followed. This personal or existential connotation of the question about a large God would be at issue for anyone who would be curious, for example, about the wager that goes with any act of following the unseen. Such a person would ask for clarification from Moses or any other prophet or mystic, any witness, anyone who claimed to be transmitting a revelation, any disclosure from a transcendent order into this limited one. We might call those mediators "the voices of experience," voices that stand behind all religions that rely on revelation or divine disclosure—which means most of them. Such prophets have their place. Had they no place, there is not much chance that the use of a term or concept such as God, in its many translations, would be at issue here at all. But the voices of experience in any immediate or direct way are not those to which the present company of scholars is responding.

How large is God? We are here left, then, not with a company of hungry hearts, thirsting souls, spirits that need renewal—although the scholars who deal with intellectual issues often may confess or demonstrate that they also are in such company. We have a company of inquiring intellects and academic passions, who come equipped with scholarly traditions, academic disciplines, and defined fields of expertise. We will return to them shortly.

A second natural question, therefore: Who are you who is doing the asking? If the asker is not the simple child or the foolish measurer without metaphor, on one hand, or the pastor or spiritual guide on the other, we have remaining the inquirer about scientia, science, and belief, also as formulated in theology. In this situation, "large" refers not to physical objects but to conceptions, ideas, imaginations, and claims. These conceptions may be expansive and indeed may elicit terms such as those that describe God as "infinite," even if infinitude implies a qualitative leap beyond any measurements where "largeness" has meaning.

How large is God? in such a scientific or philosophical case is a metaphoric question. It has to be voiced with some urgency, if the ones who ask it fear that small conceptions, confining ideas, limiting imaginations, and cramping claims about God will have bad intellectual or social consequences. What some of these consequences may be is spelled out throughout these essays. Those who pose the question in the present context have made it clear that they have an interest in helping move the God-question beyond the confines in which it is all too conventionally heard.

So when the first question, Why do you ask? remains on the agenda, it is to be answered by a threefold response: Because if you are to deal with the concept of God at all, then if God, your God, the God in your world of conception and ideas and practice, is small, you will, first, do justice to reality—and all responsible people should seek to be true, to speak the truth as well as they can. Second, you will simply be left behind in the intellectual marketplace by those who ask non-God or anti-God questions. Or, third, you will constrict the imagination and obstruct the will, at a time when our cultures need vast imaginations to deal with vast problems. They also need liberated wills, so that people in these cultures can make proper moral address to the issues of the day.

Also related to the first question is a third, which can be treated briefly here because it poses the theme that preoccupies us henceforth: Whom are you asking? Those asked are participants in intellectual and interdisciplinary forums such as this. As meaningful as the questions of piety and searching are in their contexts, they have little direct bearing on the subject when, as here, the relations of science and religious thought become the focus of concentration. We will henceforth restrict ourselves to such issues, even if they have been minority concerns in the families of faith, where theology is conventionally undertaken.

Expertise is demanded, no matter how modest the specialists may personally be. You do not consult the Answers section of an algebra text when the Questions come in a book of French grammar, although the writers of both texts may, as amateurs, know each other's field. You gain confidence from awareness that the one being consulted has, through decades of application, schooled him- or herself to address certain issues in specified ways and "has gotten good at it" or has been recognized as having done so.

Two Fundamental Contexts for Response to the Third Question

Academic disciplines and the professions organize and have organized knowledge in particular ways. People in different times and different places have called for and been responded to in diverse ways of doing this organizing. Over time in the Western world, for instance, astrology was displaced in the serious academy by astronomy; haruspication gave way to meteorology when experts wanted to forecast anything about tomorrow; alchemy found no place in the laboratories as chemistry developed.

Disciplines can become obsolete; they can be seen as having been falsely conceived in the first place. New disciplines arise with new inventions and challenges, and they keep being argued about and redefined. No cosmic referee finally decides where the lines should be drawn and no universal cartographer can or would certify that the disciplinary lines drawn in one place or time permanently match the map of reality, if such a map could be found or such reality hypothesized. But in a world of changing paradigms and intellectual conventions, scholars who want to make some sense to others do use provisional and relative maps of knowledge.

How large is God? in this book is addressed to contributor-scholars, each of whom is reasonably at home with and follows some of the conventions that go with different disciplines. They see each of these disciplines as pointing to or providing a "Source of Information." We have already met the mystic and prophet, the witness and counselor, in every case, the voice of experience, as sources of information of particular sorts. But now we deal with scientists and science and the particular sorts of sources of information they offer. Ordinarily, they deal with mathematical or other scientific formulas or symbolizations. Recall that such a division of labor follows a partly arbitrary, and no doubt provincial (Western) ordering of the map of reality, but it will do for present purposes.

The predominant sets of disciplines represented in this collection are scientific and their representatives are scientists. In what many call an "age of science," when sciences and scientists claim respect and intellectual hegemony, such predominance is in order. Recall the fact that through the centuries, all over the globe, the scientist customarily has been seen as spiritually aloof, preoccupied with other questions than those having to do with God. Since the scientist has often been under suspicion, or has been the victim of neglect by others who experience and witness to God, as well as by those who study texts devoted to witness and experience, it is fitting that, following laws of compensation and fair play and to assure that underheard voices be heard and not merely overheard, scientists should be now convoked to address how large is God?

So urgent and compelling, so promising, are the voices of scientists in our culture that they may even crowd out the witnesses of text-based disciplines such as theology and philosophy that historically have been dominant. The scientists may compensate for past neglect and in some cases set out to make the intellectual case that such theological witnesses have been guilty of presenting a too-small God. In certain kinds of gatherings and colloquies, one hears the text-people dismissed precisely on such grounds. The ancient texts with which they deal, it is said, even if these were seen as revelations of God, were written at a time when a young and small world and universe were all that were intellectually available to the imagination. Of course, it is further said, God was feared and loved as the one who was larger than the universe, who anteceded and created it, and who would bring it to its consummation and end, in those religions that spoke of endings. But all the evidence of the texts suggest that this God was still "smaller" than any that matches the picture of reality currently offered in physical and other sciences.

These ancient texts that witnessed to the immediate experience of God later proved intellectually unsatisfying because they were written by those who did not and could not conceive of the larger universe and the more expansive ideas that are characteristic of inquiry today. There have been at least two other problems associated with them, one official and authoritative, and the other unofficial and reflexive.

In respect to the former of these two: It happens that the texts that have been read as witnesses to all the reality there is, to God as the Creator, Sustainer, and Encompasser, belong to communities of interpretation. Of course, individuals in most religions, if they are literate, have had access to the texts and even where authorities would try to impose interpretations, many have given evidence that they engage in private and individual readings. Some of these readings were first interpreted heterodoxies. In the course of time, thy came to be seen as orthodoxies. They were uttered as heresies but became approved dogmas in the course of time. But all along, these communities of interpretation have tended to see the development of hierarchies and authorities that describe what is to be believed and that prescribe what is to be conceived about God. However much hierarchies and authorities might be respected for some features of life, say such scientists, they deserve to be dismissed and are dismissed if they merely invoke their authority where questions of intellectual substance are addressed. When authority thus asserts itself, it is charged, the text-interpreters who are under it must stifle their imaginations and constrict their wills. Who could think of them as being honest transmitters from sources of information about God?

The reflexive corollary to this response to authority goes along with the role of the text-keeper as custodian of a tradition. One must say in most cases in the longer past and in special instances in modern times, the keepers of the texts were assigned the roles of museum-keepers, or they fell into such roles. They then became the custodians of curios, which took the form of dogmas or other congealments. The theologians as custodians dusted the antiques in the gallery of preserved items that once had been recognized as vital. They were seen as encysted by custom, living under carapaces of convention. Such religious thinkers were unable to exert imagination or to assert will in the competition of ideas and prescriptions that go with life in a free world, in a race of progress, however that is defined and wherever it is discerned.

When the museum-keepers and interpreters today mediate such texts in such obedient and conventional ways, there are good reasons for scientists to mistrust or dismiss them. The theologians may have been seen as having had everything to offer in prescientific cultures. But they can only deal with a too-small God in a scientifically based culture and society. In ages past, one consulted them first about the size of God as the question is here posed; in the present time, one would consult them last, if at all.

We are on the verge of moving into the mental furnished apartment labeled "The Two Cultures Thesis" of the late C.P. Snow,[2] who posed scientific versus humanist cultures against each other. Aware of the ways in which these two cultures have overlapped in academies and are co-represented in the minds of most informed scientists and humanists, we can still be aware of their contours and modes in specialized ways when one poses the issue of how they deal with sources of information about the largeness of God.

Unwilling to be confined then by all implications of the Snow thesis, which deservedly has met much criticism, we still are left with the sources of information question and two sets of addresses to it. When one brings up the question of information as opposed to experience (even if it is "information about experience"), the issue is intellectual, and the intellectual schools, disciplines, and differentia come into play. Otherwise we cannot understand the language games scientists in their camps and philosophers or theologians in theirs play. If we do not keep the disciplines clear, there is no way to use terms about God with conceptual propriety; no way to communicate intelligibly across boundaries of what could appear to be incommensurable universes of discourse.

"Theology" and "Culture" as Limited Sources of Information

In the present company, we have been assigned the task of assessing how "theology" in particular and "culture" in general are sources of information about the themes that might contribute to spiritual progress, about answers to the how large is God? question. To connect the particular and the general here we appear to have to ask first, What is theology? or What do theologians do? and then What is culture? and How do theological scholars have access to and treat culture? The former pair of questions has elicited shelves full and the latter pair, libraries full, of books of answers. We need only treat themes relevant to how sciences and humanities find language to address the size of God question

Culture as such is too embracing a term to be of use for scholars treating sources of information. Emphatically, we point out and will keep on reminding ourselves, scientists are as much a part of culture as are those in other disciplines. And the other disciplines are part of the scientific culture. For now, however, we will deal with the images of culture that go with "nonscientific" disciplines as means of access.

Almost half a century ago, two scholars collated and treated then-current 164 prime definitions of culture, but they admitted that about 300 more definitions could be discovered implicitly and in disguised forms strung through their book. (Today, given almost fifty more years of scholarly controversy and disciplinary proteanism, there would be hundreds more!) They despaired of confining all these definitions within their own discipline, anthropology, or of doing justice to all to which their tentacles attached themselves. For what it is worth, A.L. Kroeber and Clyde Kluckhohn did come up with one condensation that, they said, was "close to the approximate consensus" with which they emerged from their inquiries:

> Culture consists of patterns of and for behavior acquired and transmitted by symbols, constituting the distinctive achievements of human groups, including their embodiments in artifacts; the essential core of culture consists of traditional [=historically derived and selected] ideas and especially their attached values.[3]

At once it can be seen that any such condensation or consensus suggests the relevance of cultural studies to anyone inquiring about sources of information on the size of God question. To highlight it: The talk about God has inspired patterns of and for behavior. Second, God-talk occurs within communities that respond with and inspire ritual and ceremony, taboo and prescription, manners and morals. These are acquired and transmitted by symbols. This is the case in iconic or iconoclastic, text-based and mathematically-symbolized disciplines and subcultures alike. These are all embodied, third, in artifacts such as temples, libraries, dogmas and creeds, definitions, and academies. And their essence or core is traditional [=historically derived and selected], a fourth feature that colors all that they represent as sources of information. And, fifth and sixth, these are both ideas and they have attached values.

These six dimensions may create problems for scientists who wish to contribute to spiritual progress and enlarge the picture of God for the sake of expanding the imagination and increasing the possibilities for expression through the human will. Let us revisit them: Patterns, says the scientist, are made first to be followed and then to be greeted with skepticism and possibly broken. In recent decades more than before many scientists are coming to recognize that they and their work exist within or make use of temporary paradigms and conventions. But, when serious, they would say that the scientific method(s) would call them to disrupt patterns that they find to be confining and obsolete. Would keepers of texts do so, if such guardians are representatives of traditional and authoritative communities of interpretation?

As far as behavior is concerned: There may be wisdom in dogmas and decalogues, and there can be credible authority in the commands and inspirations that prefigure conduct. But to most scientists, it belongs to the nature of their disciplines to see whether new discoveries might not force an alteration in behavior. One acts differently in the light of medical discoveries and inventions than one did before dissection of corpses or autopsies were permitted or known. Behavior patterns change when one invents hydrogen bombs as replacements for cudgels or bows. These behaviors may all trace back in the end to the how large is God? question, and scientists chafe when they are not allowed to ask their own versions of it and come up with shattering answers.

Symbols are among the chief offenders, it is said, and the roadblocks to fresh inquiry about the size of God, the conceptions that might liberate an issue in progress. For, as many scientists—all of whom use symbols—see it, traditional and historical symbols of religion serve to code and encapsulate concepts of God from earlier times, when less was known about the physical universe. The God pictured in medieval Western art was an aged and bearded character atop or slightly larger than the spheres that made up the earth and the starry heavens. And these were all rendered only slightly larger than the humans being created or impinged upon by God. The God conceived of as larger than an expanding universe in symbolization gets confined into symbols of a Bo-tree, chalices of wine and plates of bread, a stone at Mecca, or a cross at Jerusalem—all of them long ago, far away, and so small! How large is God when one is bound by such symbols?

Artifacts: nothing is likely to be a greater barrier than the artifacts, the cultural shapings and products that for all their beauty, also can confine. The Qur'an and the mosque are artifacts, as were the now acknowledged obsolete temple for Jews and the scrolls that still cause their scribes to bend over instead of looking up and out. The paraphernalia of Buddhist and Hindu worship, the relics and reliquaries of Christianity, and the congealing creeds—all these are artifacts, obvious human products. The skeptical scientist has to ask: Who says these elements square with realities of the universe whose galaxies we can peer into, whose black holes and antimatter we can hypothesize and dance around, whose genetic codes eluded even the wisest when the sacred texts were written? Artifacts almost always "reduce" the deity or the sacred they are supposed to embody, represent, guard, or help transmit.

Most problematic are "traditional=historical ideas and values." For all the poetic reach of traditional texts and historical interpretations and ideas of God and conceptions of values, they cannot do justice to the ever unfolding ideas and conceptions of the sciences of modern times. I do not mean by any of this to convey the notion that hubris, arrogance, is all on the side of scientists in their postures of skepticism and dismissal. The scientists could make out just as strong a case that it is the scribes and custodians of texts and traditions who have been arrogant in their unwillingness to see these ideas and values challenged.

Some humanists find it important to stress that scientific cultures and scientists-in-culture also produce patterns, behaviors, symbols, artifacts, ideas, and values—and may not always have tools that theologians or other humanists possess for criticism of these "products." They certainly do not have at hand such tools as those used by disruptive and irritating prophets in text traditions. I do mean that there appears to be something integral to, something structural about, the assignment of scientists in their disciplines that makes the humanistic disciplines, for all their poetry, seem problematic and inhibiting.

If all that is the case, we need a framework for conceptualizing this set of counter suspicions, to remove the issue from the field in which charges of "arrogance" and absence of "humility" may be based on emotive assessments, not intellectual grids. I will adduce two thinkers who offer conceptions for this: the late British social philosopher Ernest Gellner and the critic George Steiner. They are typical of a large cast on which we could draw. What they have to say suggests why theology and other of the humanistic disciplines are regarded problematically in recent decades, if not centuries, in respect to sources of information about the size and conceptions of God.

Gellner brought up the issue when dealing with the crisis in the humanities. Before we discuss what he has to say, it is necessary if not to define, at least to suggest what their disciplines are about. Gellner offered a condensed definition, but here is a slightly elaborated one, produced by a Commission on the Humanities after it surveyed the field between 1978 and 1980. The authors properly recognize that the humanities normally are out to define or describe the human, but, as Gellner would point out, they include the divine:

> Through the humanities we reflect on the fundamental question: what does it mean to be human? The humanities offer clues but never a complete answer. They reveal how people have tried to make moral, spiritual, and intellectual sense of a world in which irrationality, despair, loneliness, and death are as conspicuous as birth, friendship, hope, and reason. We learn how individuals or societies define the moral life and try to attain it, attempt to reconcile freedom and the responsibilities of citizenship, and express themselves artistically. The humanities do not necessarily mean humaneness, nor do they always inspire the individual with what Cicero called "incentives to noble action." But by awakening a sense of what it might be like to be someone else or to live in another time or culture, they tell us about ourselves, stretch our imagination, and enrich our experience. They increase our distinctively human potential.[4]

Those last lines suggest one kind of response to the question Why do you ask about how large is God? Since the dimension of deity and the divine by definition cannot be measured, it becomes clear that what is at stake here are the human imagination and will, which the humanities, as the commission describes them, set out to address. These humanities include philosophy, literature, comparative law, art history, religious studies and theology, ethics, linguistics, cultural anthropology, history, and more. How are they sources of information about how large is God? Why are there problems with what they are and what they do, in this context?

Here Gellner is helpful. He defined a "humanistic culture" simply as a culture based on literacy. We do not have to be literal about literacy: The humanities artifacts may include texts, monuments, choreography charts, temple floor plans, maps of cathedral cities, tapes and films. Humanistic culture "is distinguishable from illiterate 'tribal' culture on the one hand, and from more-than-literate scientific culture on the other." (Gellner acknowledged, as we must in the context of the present topic, that the term "humanist" is unfortunate. It is a survivor of the days when concerns of "human" literature were distinguished from "theological, divine concerns." But "for contemporary purposes," Gellner properly noted, "it is the literacy, and not its mundane or extra-mundane orientation, which matters. Humanist concerns now embrace the divine. (Both speak the same language.)" This is certainly the case with respect to sources of information about the size of God.

Gellner noted that the source of information question was evident in the case of believing communities responsive to particular revelations that get to be enscripted. "Humanist intellectuals, as experts on words, and above all on written words, are the natural intermediaries with the past and the future through records; . . .[and] *with the transcendent when the Word is held to contain the Message from it. . .*" [Emphasis mine] But in a world of pluralism and multiculturalism, of mass communications and easy travel, it is hard to build open cultures or free societies around response to a particular appropriation of a transcendent Word. So the role of the scribe, the theological cleric, has been redefined, constricted. The traditional way of putting this is: At the university in the medieval, official, elite, hegemonously Christian culture, theology was "the queen of sciences," and she has been dethroned subsequently.

That is not the current issue. For if theology's abdication or dethronement simply has meant that she yielded to other humanistic disciplines, wherein the claims of inscribing a message from a transcendent word is not being made, she today shares her new location, some would say her dethronement, with all those other humanities disciplines. Gellner presented a rather sardonic but not unrealistic view of the situation in a scientific culture and society.

The literate intellectuals [in a humanistic culture] become the guardians and interpreters of that which is more than transient, and sometimes its authors.

This role was one which they once fulfilled with pride. The notion of the Priest or the Scholar, or even the Clerk, evokes an image which is not without dignity; for some men and some societies, it is has more dignity than any other.

But this sense of pride is conditional on the fulfilment of the central task of this estate, which cannot but be one thing—the guardianship or the search for truth. If this is gone, only a shell remains. When the age of chivalry was over, Don Quixote was a joke. The military equipment of a knight could no longer be taken seriously. The question now is: how seriously does one now take the equipment of the clerk?

The answer is, alas: not very much. It varies a good deal, of course, with the subject matter, the milieu, the context. But giving a general answer, it is correct to say that the clerk—i.e. the literate man whose literacy led him to acquire good knowledge of the written word, an understanding of that which is written—has lost much of his standing now as a source of knowledge about the world. The educated public in developed countries turns to the scientific specialist when it wants information about some facet of the world. . . .

The deprivation of the humanist intellectual of his full cognitive status has happened fairly recently. . . . The magnitude and profundity of this social revolution can scarcely be exaggerated. There are still many members of the humanist culture who do not fully perceive what has happened. Some feel it, but angrily deny it: others feel and know it, and react with the kind of shame which needs must befall a caste when the basis of its identity and its pride has been destroyed.

Gellner found some humanists denying the charge; others were angry, or ashamed, and reactive on defensive grounds: The scientist discovers the hydrogen bomb but the humanists convey the issues hinging on them:

Hmmm. This comes close to making an advertising agency the paradigm of the republic of the mind: the statisticians find out the facts, the humanists devise slogans for persuasion. But even assuming that every humanist is a potential Persuader—what an undignified end for those who were once the Knowers!

An additional problem results from "the loss of monopoly of literacy. The clerk is a nobody not merely because he is not a scientist, but also because in the developed societies everyone is now a clerk." As a "knower," a source of information about the size of God, the do-it-yourself devotee of "spirituality" today is accorded the same prestige, maybe more, than the scribe or professor in religious studies and theology among the humanities. Assessments depend upon the charism, rhetoric, and the "persuasion factor" of entrepreneurs in the spiritual realm, and not on the knowledge about God imparted or claimed by the one who studies texts that had been sources of information in prescientific cultures.

Gellner addressed the issue of "two cultures" and saw it, as do I, as misconceived when seen as "a problem of communication between two cultures." That problem existed, but was superficial and relatively easy to solve, thought the British philosopher.

> The real and deeper problem concerns just what, if anything, it is that the humanities have to communicate. The language of the humanities is incomparably closer to what we are, to the life we live, than is the language of science; but on the other hand it is not obvious that the humanities contain, in any serious sense, genuine knowledge.

An instinctive and conscious awareness of this situation, I would contend, has been an element that has given rise to the whole enterprise of relating science to theology and religion. Typical among these are the Center for Advanced Study in Religion and Science, the Center for Theology and the Natural Sciences, Het Bezinningscentrum, the Institute on Religion in an Age of Science, the Science and Religion Forum, the John Templeton Foundation Humility Theology Information Center, and dozens more. These are not necessarily humanistic and theological centers of defensiveness and reaction, but they have been inspired by a sense of urgency to address cultural change after what Gellner called "this social revolution."

Gellner wrote before what he later called "postmodern relativism" came to characterize much of the humanities venture in much of the academic world. Whatever else it has done and however passing it may be as an intellectual movement, postmodernism, especially in its deconstructive forms, has only further subverted any notions that "the clerk" has much by way of "cognitive equipment" or "knowledge" (as opposed to "persuasion," at most) to offer. [5]

The Cognitive Equipment of the Scientist

Parallel to Gellner's statement of the situation is that of critic George Steiner whose career was devoted to the question of the culture of literacy in a scientific age. Gellner, already in 1964 , was citing an essay by Steiner that has come to be regarded as a modern classic. In "The Retreat from the Word," Steiner paid respect to some forms of Oriental metaphysics, to Buddhism and Taoism, in which "the soul is envisioned as ascending from the gross impediments of the material, through domains of insight that can be rendered by lofty and precise language, towards ever deepening silence." Therefore, "the ineffable lies beyond the frontiers of the word." But these are non-God religions. Their insights are relevant to discussions of spirituality and religiousness but not to God-centered world views and religions, in which theologians (theos+logos, those who have a "word" about "God") are represented. The thinkers in these traditions may have much to say about the science and religion issue; it is no wonder that some have spoken of the Tao of physics. But the non-God religions by definition and without complaint about exclusion cannot speak to how large is God? Their inquiries and eventual or final silence does not have to do with God of any size, metaphor, or analogy.

"The primacy of the word," however, "of that which can be spoken and communicated in discourse," said Steiner, was a theme characteristic of the Greek and Judaic genius and carried over into Christianity.

> The classic and the Christian sense of the world strive to order reality within the governance of language. Literature, philosophy, theology, law, the arts of history, are endeavours to enclose

within the bounds of rational discourse the sum of human experience, its recorded past, its present condition and future expectations. [The classics of these traditions] bear solemn witness to the
belief that all truth and realness—with the exception of a small,
queer margin at the very top—can be housed inside the walls of
language.

Steiner knew and contended that this belief is no longer universal
and has been in decline in the West since the age of Milton. He then
proposed a theme of great importance for our topic of sources of information. The formulation of analytical geometry and the theory of
algebraic functions, of calculus, in the age of Newton and Leibniz,
Steiner noted, led mathematics no longer "to be a dependent notation, an instrument of the empirical." Instead, "it becomes a fantasti-
cally rich, complex and dynamic language. And the history of that language is one of progressive untranslatability." Between verbal
languages there can be translations, as from Chinese ideogram to
English paraphrase. But mathematical language is self-contained, not
translatable to verbal language. "This is a fact of tremendous implication. It has divided the experience and perception of reality into separate domains."

> . . . a branch of inquiry passes from pre-science into science when
> it can be mathematically organized. It is the development within
> itself of formulaic and statistical means that gives to a science its
> dynamic possibilities. The tools of mathematical analysis trans-
> formed chemistry and physics from alchemy to the predictive sci-
> ences that they now are. By virtue of mathematics, the stars move
> out of mythology into the astronomer's table. And as mathemat-
> ics settles into the marrow of a science, the concepts of that sci-
> ence, its habits of invention and understanding, become steadily
> less reducible to those of common language.

Passingly, Steiner noted, "It is this extension of mathematics over
great areas of thought and action that broke western consciousness
into what C. P. Snow calls 'the two cultures.'" As late as the time of
Johann Goethe and Alexander von Humboldt, exceptional people
could feel at home in both mathematical and humanistic cultures.

Steiner commented: "This is no longer a real possibility." Here Steiner sounded like Gellner:

> Except in moments of bleak clarity, we do not yet act as if this were true. We continue to assume that humanistic authority, the sphere of the word, is predominant. . . [Today] all evidence suggests that the shapes of reality are mathematical, that integral and differential calculus are the alphabet of just perception. The humanist today is in the position of those tenacious, aggrieved spirits who continued to envision the earth as a flat table after it had been circumnavigated, or who persisted in believing in occult propulsive energies after Newton had formulated the laws of motion and inertia. . . . It is no paradox to assert that in cardinal respects reality now begins outside verbal language. . . .
>
> Few humanists are aware of the scope and nature of this great change. . . . Nevertheless, many of the traditional humanistic disciplines have shown a deep malaise, a nervous, complex recognition of the exactions and triumphs of mathematics and the natural sciences.

Steiner showed how historians; economists after the period of the classic masters like humanists Smith, Ricardo, Malthus, and Marshall ["econometrics is gaining on economics"]; sociologists; even philosophers were retreating from the word and aspiring to be exact scientists, measurers, and numberers. One might add that the introduction of "rational choice theory," as the all-purpose explanation of religious response, is still another evidence of this mathematical impulse thrust into the realm of the spirit. 6

Just because Gellner, Steiner, and a host of less articulate witnesses that could be adduced have said such things does not make them true. But they can be tested in numberless experiments, from those wherein one assesses the curricular choices of university students through observations of the incomes paid investment counselors, the awe shown self-deprecating "numbers crunchers" in the social sciences, the necessity for using mathematical language to dominate in medical and other scientific research, and to the equations of astrophysicists, while poets of the stars suffer neglect as sources of information. It becomes clear that the "cognitive equipment of the clerk" is perceived dimly and ambiguously by serious publics in the scientific age, if, indeed, expectations for more than rhetoric from the clerk show up at all.

All this has a bearing on the issue of sources of information and about the answers to the how large is God question. It may also go a long way toward explaining why the scholars at various centers and consortia on science and religion often complain that, while they have more precise and "bigger" answers to questions of heart and soul and spirit than do those we might call theologians of existentialism and personal faith, and while their canvases ("the sciences in a scientific age") are vast, they remain rather small elites talking to specialists.

The publics, meanwhile, when not concerned with sources of information other than reports on experience, turn to the patterns and behaviors, the symbols and artifacts, the traditions and ideas and values of those who make more conventional and less scientific-mathematical statements about the size of God than do scientists. Such publics, uncertain about the cognitive equipment of the clerk [=cleric], still seem relatively content to hear that God is large enough to guide their own personal destinies, and to be with them in the intimate spheres where, according to Pascal, "the heart's reasons, that reason does not know" get relied upon and expressed. In a way, the scientist as religionist has won the war but seldom wins the battles, while the existential religionist wins battles for the individual heart but not the war for credibility in an age and culture of science.

Conceptions that Show Promise

If one accepts Gellner's contention that what is at stake is communication between two cultures, the problems about how large is God questions would be on the table, perceived as difficult, but still in the range of solubility. If one goes on to accept Gellner's and Steiner's (and others') argument that mathematical language is untranslatable into verbal languages, and vice versa, and sees nothing but the isolation of the humanist (including theologian) "clerk" who has been deprived of "cognitive equipment" and rendered unable to address issues of "knowledge," there is no point in including science and scientists in the convocations of humanistic scholars, including theologians, who on entirely different terms discuss sources of information about God and the size of God. Conversely, there would be no point in bringing up "theology and culture," as here defined in respect to the humanities, in a forum like this where scientists dominate. And we would consequently leave two publics or two sets of public hungers similarly separated and isolated from each other, thus contributing to cultural and social malaise and gloom.

Some compensatory or bridging ideas and words, however, are available, and they merit at least preliminary inquiry and discussion. Gellner offered one such clue when he said that "the language of the humanities is incomparably closer to what we are, to the life we live, than is the language of science," even though, as we have read and I have concurred, "it is not obvious that the humanities contain, in any serious sense, genuine knowledge." That acknowledgment of a position and a role for the humanist clerk was not presented by Gellner as an act designed to retrieve dignity for the beleaguered text-person, although the British philosopher was capable of such acts in other contexts, wherever he discussed social philosophy.

It was Steiner who in the present context did go further in retrieving the role of the humanist scholar, although still under the shadow of the hegemony of mathematics or science. Steiner turned to a scientist, J. Robert Oppenheimer, who at least held out some "small hope" for poetry, the humanities, for "historical, moral and social inquiry in which the word should still be master," in the face of the "strident muteness of the arts" in his time, our time.

> Oppenheimer began by pointing out: that the breakdown of communication is as grave within the sciences as it is between sciences and humanities. The physicist and mathematician proceed in a growing measure of mutual incomprehension. The biologist and the astronomer look on each other's work across a gap of silence. Everywhere, knowledge is splintering into intense specialization, guarded by technical languages fewer and fewer of which can be mastered by any individual mind.[7]

We who participate in the writing of this book have seen evidences of the difficulty of creating mutual comprehension, of being able to express empathy, and of communicating intellectually around the table of the humility theology panels, where participants have strong commitments to engage in a common enterprise and much respect for each other. Little wonder that "back home" in their several universities or professional societies, they experience even greater difficulty getting a hearing.

Did Steiner, do we, call this difficulty a sign of hope? It is a beginning because it calls forth humility and modesty of claim, also about sources of information in science, about the size of God. What is needed, suggested Oppenheimer as Steiner cited him, "is a harsh modesty, an affirmation that common men cannot, in fact, understand most things and that the realities of which even a highly trained intellect has cognizance are few."

Steiner did say that the sombre view just mentioned "seems unassailable" in the sciences.

> But we should not readily accede to it in history, ethics, economics or the analysis and formulation of social and political conduct. Here literacy must reaffirm its authority against jargon. I do not know whether this can be done, but the stakes are high. [After illustrating from the language of politics and media in our time, he wrote that if the trend continues O]ur lives will draw nearer to chaos. There will then come to pass a new dark age. The prospect is not remote: . . . [8]

To Oppenheimer's (and here Steiner's) first word about modesty we might add a second based on observation. The reader will note that the scientists in a colloquium such as this, after bringing the mathematical language and its associated "cognitive equipment" to bear on the how large is God question, also turn into humanists or express themselves in forms congenial to the humanities readership when they apply it. They are then, in all the essays and books on this subject, rhetoricians, persuaders, historians, philosophers; some of their friends and all their enemies would say theologians, and not simply "pure" scientists. In this sense, whatever they do in their laboratories or observatories, in the interdisciplinary forums they join the company of Eugen Rosenstock-Huessy, who was uneasy about a culture of pure scientific objectivity and in an essay "Farewell to Descartes" told why he had to write in two modes:

> I am an impure thinker. I am hurt, swayed, shaken, elated, disillusioned, shocked, comforted, and I have to transmit my mental experiences lest I die. And although I may die. To write a book is no luxury. It is a means of survival. By writing a book, a man frees his mind from an overwhelming impression. The test for a book is its lack of arbitrariness, the fact that it had to be done in order to clear the road for further life and work.[9]

Could it be that the very impulse to want to translate the "mathematical" language of science into the verbal language of the humanities is moved by such impulses on the part of scientists as they experience wonder or mystery—and seek appropriate modes for expressing it? Much of the finest humanities and sometimes theological writing of our day comes from venturesome first-rate scientists who risk "impurity" of expression and violation of disciplinary boundaries in order to speak on urgent issues, including the how large is God question.

I have often noticed that more scientists are good at humanistic expressions than humanities experts are at scientific statement. It may be that the scientists engaging in the act of transition from mathematical language (and all its analogues) to verbal language are reminding themselves and showing us that they know, in Gellner's words, that "the language of the humanities is incomparably closer to what we are, to the life we live, than is the language of science." This may be illustrated by reference to the ways publics have of relying on technological "mathematical" medical science for some things relating to their bodies, while drawing at least supplementarily on alternative medicine, holistic medicine, and, of course, meditation and prayer, to address other aspects of their being.

A Renewed Place for "Humility"

Similar modesty about those in the humanities and theology is equally in place. I have tried to show why that is the case and to mention the demoralization or revision of program among many humanists, including theologians, so little more need be said here about the arrogance of transgressors. But exhortations to modesty do not deal substantively with the issue before us. Acts of retrieval and restoration, modest as they are, can proceed further than mere exhortation.

For one instance, while the culture and society of a republic cannot be officially founded on and grounded in a particular grasp of a revelation of the logos, which would mean "the word," it is true that in the voluntary and associational side of republican life, there is a very widespread sense that these ancient texts, now as always, serve as a kind of source of information (indeed, the basic, the only one) for access to "the transcendent when the Word is held to contain the Message from it." This is the case with those, also in this scientific cultures, who are enduringly and even newly responsive to the Qur'an, the Torah, the New Testament, or any number of ancient holy books. To their prophets and potentially to their adherents, these are not humanities texts about dead realities and a dead or, maybe worse, too-small God.

The problem before us now is to suggest how those who use ancient texts, which claimed to be sources of information about God, can increase imagination and enlarge the scope of the will (as expressed in moral decision-making). In the humanities communities, the scholars and articulators, if they are to serve the source of information about how large is God, are engaged these years in tasks of new inquiry about the use of the texts.

Philosopher Paul Ricoeur[10] has paid respects to the limited function such texts play when the "clerks" are historians and only historians. Believing publics do turn to them, as biblical scholars and commentators, for what turns up, thanks to their cognitive equipment. By studying the ancient contexts of the logos, they offer kinds of impartation that can serve the imagination and will. They are therewith potentially able to relate to a "larger" God and to more encompassing realities than are those who are confined by museum-shopkeeping approaches to history. So historical inquiry of the sort in which theologians engage need not be seen as looking only backward, even though, in Ricoeur's words, it can do no more than suggest "the world behind the text."

Second, what if the scholars are literary critics? There is the "world of the text," Ricoeur's tribute to another specialist, the literary scholar. Such a person contemplates how large is God in the formal testimony of psalms, cantos, suras, parables, epistles, enigmas, mysteries, puzzles, or preachments. The forms have much to do with what is disclosd. Scholars who study these also serve in a limited way, by pointing to the text as a source of information.

The third, for Ricoeur, is "the world in front of the text." Here the humanistic scholar turns ethicist, moralist, poet—or invites others to turn thus—for the sake of envisioning ways of life that would otherwise not have been entertained. Thus, the world in front of the text is not dogmatically bordered, stifled by creeds, hierarchically limited, or on authoritarian grounds obscured from access. Instead, it serves as a source of information for the release of imagination and the stimulus of will so that creative and moral action can follow, and the intellect can be open for new possibilities—such as those presented in the sciences.

When such understandings appear, the theologian will not be so readily circumscribed, sent to the end of the table, or seen only as "persuader," even if that theologian is not restored to regal status with the Queen of Sciences, or recognized with full comprehension of and respect for her cognitive equipment. But to advance the enterprise, there must be great care taken in the use of the languages and modes of discourse. Let me propose one provisional framework, useful for others who might devise their own for dealing with these modes.

The Modes of Inquiring about How Large Is God?

British philosopher Michael Oakeshott, in a precocious and still, in my view, undernoticed book and life work, provided an analysis of Experience and Its Modes that is helpful here. When the scientist speaks of the size of God, he or she operates within the context of one or another of "provinces of meaning," "universes of discourse," or, for Oakeshott, "modes of experience." Experience, Oakeshott argued, is a whole. A mode is a consistent, particular way of viewing the universe, of experience; within it one directs attention in a particular way. Every idea, including every idea about the size of God, belongs to one or another mode. "An idea cannot serve two worlds." To forget this is to be prone to ignoratio elenchi, which means, as scholastics of old would say, making a "category-mistake." Philosopher Oakeshott could be quite astringent: "The sign of all seriously undertaken thought" is its "freedom from extraneous purpose and irrelevant interest."

Theoretically, there could be an infinite number of modes, but Oakeshott elaborated on only three or four basics ones, all of them relevant here. He reviewed inquiry in each under some category or species designation. Thus, years after he wrote Experience and Its Modes, he spoke of Art, which could be viewed—my term, not his— as viewing life sub specie imaginationis, in the light of the imagination. History, qua history, restricts itself to organizing the totality of experience sub specie praeteritorum, in the light of the past and nothing but the past. When historians use historical data as a source of information about the future, they necessarily slip out of the category and adopt a mixed mode, which is useless, or another mode, which is appropriate for other expressions.

The reader of the history, be it a biblical text or an inspiring story of discovery or sainthood, may move the information acquired there into another mode, that of the world in front of the text, but is not then doing history. When I move from classroom into chapel and profess to impart something of the transcendent through a Word containing the Message, I am doing something utterly different than when in the classroom, dealing with the same texts. Historians are sometimes downgraded or excluded from the table of the scientists because they cannot impart information about the future, information to which the scientist through experiment is moving. But one can help unfold the texts that contain so much possibility for enlarging the conception of God.

A third mode, most relevant here, is the scientific. The world of science is conceived sub specie quantitatis, in the light of quantity. Let me stress at once that "quantity" is not used here in a naive sense, as if science were nothing more than measuring distance to stars, the heat of specimens in the laboratory, or anything else confined to computer assessments. The words "stable, common, and communicable" in the passage soon to be quoted, may make it seem so. But they need not confine the inquiry. "Quantity" here has a kind of metaphoric character that is not removed when, as in postmodern scientific inquiry, there is stress on relativity, chaos, and instability in all reckonings.

Oakeshott was anticipating what Gellner and Steiner reduced to the language of mathematics, but here he cleared the air for any who thought that when they used the word mathematics, they meant only a circumscribed discipline.

> Science is an attempt to discover and elucidate a world of ideas before all else stable, common, and communicable. And this general end involves science in, and confines it to, the elucidation of a world of quantitative conceptions. Whatever cannot be conceived quantitatively cannot belong to scientific knowledge. The world of science is a uniform world of quantitative ideas, a world, that is, of quantitative generalizations. . . . This does not, of course, mean that the ideas of science must always be figures, or that its inferences and generalizations will always be expressed in mathematical symbols. It means that its conceptions must be subject to judgments of "more" and "less," "increase" and "decrease," " intensity," "size" and "duration."

The humanist who reads that begins to understand why a table of scientists does not laugh to hear the question How large is God since humanists have long reduced that to a metaphorical issue. Or scientists have derided anthropomorphs who make God into a big other-than-human, but comprehensible to humans. One might say that it is easier for scientists to speak of How large is God? than to ask, Is God? or to give voice to the syllable, God.

Fourth, there is the practical mode, which views experience and the world sub specie voluntatis or sub specie moris in the light of the volition or morality:

> . . . . the differentia of practice is the alteration of existence; that . . . this implies a felt discrepancy between "what is" and what we desire shall be, it implies the idea of a "to be" which is "not yet." . . . Practice as activity presupposes, it appears, two worlds which

are somehow to be reduced to one. It presupposes a present world, . . . of . . ."what is" or "existence"; and beside this, another world, so far represented as a mere "to be" or "not yet." . . . Practice is action, the alteration of existence.

Relevant to the present inquiry is Oakeshott's summation of practical activity and practice as a mode. "Religion is not a particular form of practical experience; it is merely practical experience at its fullest. Wherever practice is least reserved, least hindered by extraneous interests, least confused by what it does not need, wherever it is most nearly at one with itself and homogeneous, at that point it becomes religion. . . . [R]eligion is practical activity, and religious experience is practical experience; . . . in religion practical experience realizes its full character, religion is the consummation of practice."

Oakeshott constantly warned against "category mistakes" in respect to the modes. Thus, the historian qua historian cannot speak authoritatively about the alteration of the world in the future; he or she has no expertise in it congruent with anything drawn from disciplined work. One contingent element or event in the next moment negates or refutes the envisionings of the most learned historian a moment before. The scientist qua scientist makes a category mistake if he makes a doctrinal or theological change on the basis of quantity. If nine out of ten Catholic women have a view of birth control that contradicts official Vatican teaching, the social scientist may suggest to the Pope that he reappraise the teaching. But nothing in the quantity, the numbers, gives him authority or competence for changing a teaching. Saying that God is "large" has no practical consequence except sub specie voluntatis, a different mode of speaking. [11]

Conversation as a Means of Profiting from the Modes

Individuals, "impure" scholars a la philosopher Rosenstock-Huessy, or full human beings do not make category mistakes by being full human beings. That is, they apprehend experience which is a whole through one mode or one part of one's being at one moment and another at another. There must simply be an awareness, no doubt soon relegated or graduated to the reflective, of which mode at any time and for a specific purpose is appropriate. But what of a company made up of physicists, mathematicians, historians, astronomers, theologians, and biologists who converge on the single topic, How large is God?

Oakeshott and many others—theologian David Tracy most influentially in my case—use a "conversational" model for interaction. The world itself is "conversation" to the British philosopher. This is true of politics and education; both thrive if there is "an unrehearsed intellectual adventure" that relies on spontaneous enjoyment for expression. Conversing has something in common with love and friendship, art and play. Those who enter the game may bring conviction, but they also become aware of their own relativity in the face of the other. No one surrenders the claim or demand of integrity but only of exclusive validity. Here life is actually enacted, lived sub specie moris, where there is "a genuine and unqualified recognition of other selves." [12]

When addresses to the question of the size of God are monopolized by the historian, who can only talk of past language; by the artist, who can only imagine and express the symbols while creating something new; by the scientist, who measures and quantifies, even if figuratively, the issue of the size of God; and the person in the practical mode, who is out to change the world, they find that their conversation, not a source of information that would promote progress, is potentially part of a process of disclosure. This disclosure, this revelation, relates not only the self to the self but also the other to the selves in community, and of them to each other. The representatives of these "arrests of experience"—Oakeshott's terms—stand the chance of having achieved something of what a "large" picture of God would have done. They would have been religious and have provided new material for theological interpretation. They would have stimulated life lived "in the light of imagination" and "in the light of the will to change."

Conclusion

In this essay, I have not accepted C.P. Snow's definition of "two cultures," the scientific and humanistic, as a problem of communication but as signaling something much deeper, something that needs to be conceived in other terms than those he used. Those terms have included the notion of "the cognitive equipment" of the clerk, the humanist, and of the scientist, especially in respect to what they relate to and bring as sources of information about the size of God, how large is the God-concept. Third, we have suggested that these two, and others, can best be regarded as "modes" of experience, "arrests of experience," each with different intentions and rules of the game.

Having granted Ernest Gellner his point that when it comes to sources of information about anything, the clerk, the humanist, or the text-expert is less looked at today than in some other cultures, it may seem that I have left him or her little role in the religious fields. But once one introduces the concept of modes, a good deal of retrieval goes on. How should one conceive of this?

It has occurred to me, ever since this assignment, that the basic question to all the contributors of this book is not How large is God? in any real, answerable sense. Instead, it is, "How rich are the imaginations and wills of the people who deal with God, or the concept of God?" Here is where humility enters the picture for all the disciplines. Those who have been critical of theologians, historians, philosophers, and other humanists have, in most cases, been criticizing them for confining their imaginations in dated models of talking about God or of being obedient to hierarchies or habits that do not let their imaginations go unfettered.

The humanist has as many options as the scientist when it comes to the imagination. When confronted by those who think that "everything has been tried," so the humanist scholar must merely reproduce accurately the tried and perhaps true, the humanist might well take out a chess encyclopedia. Thus:

> **Openings, Number of Possible.** . . . [There are] 400 different possible positions after one move each. In 1895 C. Flye St. Marie calculated that 71,852 different legal positions were possible after two moves by each player. This figure is geometrically correct but [in another way of calculating] 72,084 different positions are possible.
>
> After three moves each more than 9,000,000 positions are possible. To arrive at every one of the possible positions after four moves each at the rate of one position a minute day and

night would take a player 600,000 years. There are 2x10 [to the 43 power] possible different legal positions on a chess-board, and it has been estimated that the number of distinct 40-move games is 25x10 [to the 115th power], far greater than the estimated number of electrons in the universe (10 [to the 79th power]).[13]

I once mentioned this to a mathematician, who concurred in the calculation. Then I said, "In other words, infinite!" He shouted back, "Oh, no, no, no. Not infinite. Just a very large number!" He went on to say that people in the mathematics building left the Infinite to us in the Divinity School.

If one chessboard offers that many options, one thinks of what is available to the imagination when all the cells of all the brains of all the humanists and scientists and ordinary believers and nonbelievers are put to work: not everything has been thought of or tried. What matters, therefore, when speaking of how large is God is to remember that there need be different answers depending upon the nature of the questions asked and the modes, or "arrests of experience," of the whole that are called at any moment. The largeness of God for the heart is usually expressed in personal (I-thou) terms, with what Pascal called l'esprit de finesse and the largeness of God for the mind is usually expressed in mathematical and impersonal (I-it) terms, with l'esprit de geometrique put to work.

In which is God larger? The answer to questions about the size of God in I-thou circumstances differs from those in I-it situations. Another way to put this: The question about sources of information leads us to see that "to inform" means something different to the scientist working sub specie quantitatis and the humanist=clerk=theologian=philosopher=historian=spiritual guide working sub specie imaginationis, praeteritorum, voluntatis, and moris— in the light of the imagination, the past, the will, and custom.

In such a view, the theologian is not by definition (although may be by "custom") ruled out when the largeness of God is the topic. Nor are historians, even though they deal with that past that the scientists might leave behind in their role as scientists. Steiner, again, in another essay:

Even at the sharpest edge of autistic engagement, the scientist is oriented toward the future, with what it contains of morning light and positive chance. . . . This, again, is a social and psychological estrangement to which we pay too little heed. Today's high-school student solves equations inaccessible to Newton or to Gauss; an undergraduate biologist could instruct Darwin. Almost the exact contrary holds true for the humanities. . . . The humanist is a rememberer. He walks, as does one troupe of the accursed in Dante's Inferno, with his head twisted backward. He lurches indifferent into tomorrow.[14]

That same humanist clerk, that text-oriented person, however, in religion may be called to see reality sub specie voluntatis and set out to change the world in the light of the text. I am well aware that some defenders of tradition do not see any such possibility. Thus, Edward Shils, in his major work *Tradition* by page four, footnote two, is trying to dismiss alternatives:

Those who argue for tradition nowadays turn out . . . to be very progressivistic in the traditions they support; it is the "tradition of change" which they praised and do not think about what is involved in this paradox. More substantive traditions, traditions which maintain the received, receive less support. The "tradition of traditionality" has very few supporters.

He footnoted this with "See, for one example among many others, Martin Marty, 'Tradition, verb (rare),' The University of Chicago Record, 9, no. 4 (21 September 1975); 136-138.[15] Being thus implicated, I was drawn into several conversations with my late colleague, never fully convincing him but at least alerting him to a different use of tradition when another "mode" of discourse came into play. The fundamental tradition of change in the Western world, insofar as it has been influenced by Jewish and Christian traditions—which means very much— is the prophetic. The Hebrew prophets were in this tradition. They took inherited ideas of the covenant, "the 'tradition of traditionality'" and used it in efforts to seek to effect change in the will of the king, the judges, the elders and scribes, and the people.

This approach to the clerk's texts relates to Max Weber's distinction between the use of texts and past events in the hands of the virtuoso and the charismatic figure. Michael Hill condensed this:

> In contrast with [the] rigorous restatement of an existing religious tradition, charisma represents a shattering of what exists already and the articulation of an entirely new basis of normative obligation. It aims not at pursuing ethical ideals "to the letter"- which is a characteristic claim of monastic groups—since to the member of a charismatic movement "the letter" has been effectively destroyed. While the typical statement of a charismatic leader can be given as "It is written, but I say unto you. . . .". The characteristic statement of a virtuoso is "It is written, and I insist. . . ."[16]

It takes a large God to empower the virtuoso agent of change. One thinks of Martin Luther King who, as historical theologian, would have had to deal with accurate preservation and rendering of old texts, but as prophet took the Declaration of Independence and the Constitution from one pocket and the writings of the Hebrew prophets and those about Jesus in the other and waved them, in the face of small-God immoralists: "It is written, and I insist. . . ." At such a moment, in such a kairos, no one cared about the largeness of God in science, sub specie quantitatis. What mattered, for that particular "arrest of experience" was the largeness of God to address the unimaginative, cramped, intransigent human will. In such a circumstance, people looked for and to a different source of information than they do when acting as scientists or in dialogue with sciences. Paul Ricoeur relates the "closedness" of a text to those who are traditionalist, to the "openness" of a text when it is voiced and risked in conversation:

> We can, as readers, remain in the suspense of the text, treating it as a wordless and authorless object; in this case, we explain the text in terms of its internal relations, its structure. On the other hand, we can lift the suspense and fulfil the text in speech, restoring it to living communication; in this case, we interpret the text.[17]

In such instances, there have to be at least as many imaginative openings as one chess board offers—in other words, not infinite, but "a very large number" that have not been entertained in the light of past witness, experience, story, and statement. What is at stake in all the conversations marked by "humility" in theology and science is an awareness of the modes of inquiry and discourse, a mutual respect across the disciplines, and the possibility of what St. Benedict called a conversatio morum—a lifelong changing in behavior—to match the conversatio intellectus—a lifelong changing in intellect—that the consciences of the scientists and the theologians or cultural analysts in their company demand, if they are ready to conceive of a "large" God.

Notes

1. J.B. Phillips, *Your God Is Too Small* (New York: Macmillan, 1952).
2. C.P. Snow *The Two Cultures and the Scientific Revolution* (New York: Cambridge University Press, 1959).
3. A.L. Kroeber and Clyde Kluckhohn, *Culture: A Critical Review of Concepts and Definitions* (New York: Vintage, 1963), p. 66. It had originally been published in 1952 as Vol. XLVII—No. 1 of the Papers of the Peabody Museum of American Archaeology and Ethnology, Harvard University.
4. "Report of the Commission on the Humanities," The Humanities in American Life (Berkeley, Calif.: University of California Press, 1980), p. 1.
5. Ernest Gellner, "Crisis in Humanities and the Mainstream of Philosophy," in J.H. Plumb, ed., *Crisis in the Humanities* (Baltimore: Penguin, 1964), pp. 71-79.
6. George Steiner, *George Steiner: A Reader* (New York: Oxford University Press, 1984), pp. 283-288. The essay "The Retreat from the Word" was first published in 1961.
7. Ibid., 304.
8. Ibid., pp. 303-304.
9. Eugen Rosenstock-Huessy, *I Am an Impure Thinker* (Norwich, Vt.: Argo, 1970); the essay first appeared in 1938.
10. See this elaborated in "Narrative: 'the Substance' of Things Hoped for," chapter 5 in Kevin J. Vanhoozer, *Biblical Narrative in the Philosophy of Paul Ricoeur: A Study in Hermeneutics and Theology* (Cambridge: Cambridge University Press, 1990), pp. 86-115.
11. Michael Oakeshott, *Experience and Its Modes* (Cambridge: Cambridge University Press, 1933), pp 327, 311 fn. 111, 176, 221, 259-60, 292.

12. Ibid., 197-98, 201-02; Michael Oakeshott, *On History and Other Essays* (Oxford: Oxford University Press, 1983), p. 11; Michael Oakeshott, *Rationalism in Politics and Other Essays* (London: Methuen, 1962), p. 210.
13. David Hooper and Kenneth Whyld, *The Oxford Companion to Chess* (New York: Oxford University Press, 1984), p. 232.
14. Steiner, *A Reader,* p. 198.
15. Edward Shils, *Tradition* (Chicago: University of Chicago Press, 1981), p. 4.
16. Michael Hill, *The Religious Order* (London: Heinemann, 1973), p. 2.
17. Paul Ricoeur, *Hermeneutics and the Human Sciences: Essays on Language, Action and Interpretation,* ed. John B. Thompson. (Cambridge: Cambridge University Press, 1981) p. 15.

Frontiers and Limits of Science

John D. Barrow

An apocryphal story is often told of the patent office whose director made an application for his office to be closed at the end of the nineteenth century because all important discoveries had already been made.[1] Such a fable nicely illustrates the air of overconfidence in human capabilities that seems to attend the end of every century. In science, this confidence often manifests itself in expectations that our study of some branch of nature will soon be completed. Typically, there is great confidence in the scope of a successful line of enquiry, so much so that it is expected to solve all problems within its encompass. But the quest to complete its agenda often uncovers a fundamental barrier to its completion—an "impossibility" theorem.

I want to draw attention to some of the barriers to scientific progress that may be encountered in the future—some of which may even have been encountered already. They are interesting, not because of some satisfaction at seeing science circumscribed, but because they involve key ideas in science that are as important as new discoveries. As a particular example, I will focus the discussion primarily upon the search for a "theory of everything," by which particle physicists mean simply a unified description of the laws governing the fundamental forces of nature.[2] I will consider four general types of limitation that could prevent the completion of our investigations into the form of a theory of everything.

Existential Limits

The first question is whether such a theory of everything exists at all. It is possible that some of our research programs are directed at discovering theoretical structures that do not exist. At root, this is just the old philosophical problem of distinguishing between knowledge of the world and knowledge of our mental models of the world. Limitations upon our abilities to understand fully the latter might be best interpreted as limits of scientists rather than limits of science. Bearing this distinction in mind, we should be sensitive to the sources of some of our favorite scientific concepts.

The notion that it is more desirable to seek a description of the universe in terms of a single force law, rather than in terms of two, three, or four forces, is, at root, like the entire concept of a law of nature, a perspective that has clearly definable religious origins.[3] We have a natural sensitivity for pattern recognition and for the appreciation of symmetry that quickly picks up subtle deviations from perfect symmetry. There are interesting evolutionary reasons why we might be expected to have developed acute appreciation for symmetry: It is a way of distinguishing living and nonliving things when viewing a crowded scene.[4]

Although there has been growing interest in the structure of superstring and conformal field theories as candidates for a theory of everything,[5] these have proven too difficult to solve. Hence, there are as yet no predictions, observations, or tests to decide whether or not this line of inquiry is in accord with the structure of the world, rather than merely a new branch of pure mathematics.

There have been some studies of particle physics theories in which there is no unification of the fundamental forces of nature at very high energies. So-called "chaotic gauge theories"[6] assumed that there were no symmetries at all at very high energy; symmetries emerged only in the low-temperature limit of the theory, a limit that necessarily described the world in which atom-based life forms live. These ideas have not yet been thoroughly investigated, but they are similar in spirit to some of the recent quantum cosmological studies on the nature of time. They argue that time is a concept that only emerges in the low-temperature limit of a quantum cosmological theory, when the universe has expanded to a large size compared with the Planck length scale.[7]

Even if we concede that there is a theory of everything to be found, it should not be assumed that it is logically unique. When the study of superstring theories began in the early 1980s, it appeared that there might be just two of these theories from which to choose.[8] At first in these theories, logical self-consistency appeared to be a much more powerful restriction than previously had been imagined. As investigations have proceeded, a new perspective has emerged: There may be many thousands of possible string theories. While it may ultimately transpire that these theories are not as different as they first appear, many of them may turn out to be just different ways of representing the same underlying theory. If they are distinct, then the message may be that there exist many different self-consistent theories of everything. Some of them may contain the four forces of nature that we observe, while others may lack some of these forces or contain additional forces.

While we know that we inhabit a world described by a system of forces that permits the existence and persistence of stable complex systems—of which DNA-based "life" is a particular example—if other systems of forces also permitted complex life to exist, then we would need to explain why one logically consistent theory of everything is to be chosen rather than another.

Conceptual Limits

If a deep theory of everything does exist, then how confident should we be about our ability to comprehend it? This depends upon how deep a structure it is. We could imagine an infinitely deep sequence of structures that we could only ever partially fathom. Alternatively, the theory of everything may lie only slightly below the surface of appearances and be well within our grasp to comprehend. It does not follow that the most fundamental physical laws need be the deepest and most logically complicated aspects of the universal structure.

In practice, we have learned that the outcomes of the laws of nature are invariably far more complicated than the laws themselves because they do not have to possess the same symmetries as the underlying laws.[9] However, we must appreciate that the human brain has evolved its repertoire of conceptual and analytical abilities in response to the specific challenges posed by the tropical savannah environments in which our ancient ancestors developed over half a million years ago.[10] There would seem to be no evolutionary need for an ability to understand elementary particle physics, black holes, and the ultimate laws of nature. Indeed, it is not even clear that something as simple as rationality was selected in the evolutionary process.

This pessimistic expectation could be avoided if it is true that the laws of nature can be understood in full detail by a combination of very elementary concepts, such as counting, cause-and-effect, symmetry, and so on—conceptions that do seem to have adaptive survival value. In that case, our scientific ability should be seen as a by-product of adaptations to environmental challenges that may no longer exist or that are dealt with in other ways following the emergence of consciousness. Moreover, much of our most elementary intuition for number and quantity may be a by-product of our linguistic instincts. It might even be that the structure of counting systems in many primitive and ancient cultures, together with notions like the place-value notation, derive from our complex genetic programming for language acquisition.[11] Our linguistic abilities are far more impressive than our mathematical abilities, both in their complexity and their universality among humans of all races.

We might ask whether a theory of everything will be mathematical. All our scientific studies of the universe assume that it is well described by mathematical structures. Indeed, some would say much more: that the universe is a mathematical structure.[12] Is this really a presumption? We can think of mathematics as being the description (or the collection) of all possible patterns. Some of these patterns have physical manifestations, while others are more abstract.

Defined in this way, we can see that the existence of mathematics is inevitable in universes that possess structure and pattern of any sort. In particular, if life exists, then pattern must exist, and so must mathematics. There is no reason to believe that there exists any type of structure that could not be described by mathematics. But this does not mean that the application of mathematics to all structures will prove fruitful. Indeed, the point is tautological: Given another type of description, this would simply be added to the body of what we call mathematics. Physical science has become defined by a history of successful applications of mathematics to the structure of some aspects of the world. Physical science owes its success to limiting its objectives to those problems in which mathematics does more than merely describe in a succinct fashion. Sometimes this success occurs through current mathematics, but occasionally new types of mathematical modelling must be developed to deal with a new type of pattern. For example, the development of statistics to deal with trends in large samples of data.

Let us return to the issue of our evolutionary development, and the debt we owe to it. One way to look at the evolutionary process is as a means by which complex ("living") things produce internal models, or bodily representations, of parts of the environment. Some aspects of the astronomical universe—like its vast age and size—are necessary prerequisites for the existence of life in the universe. Billions of years are required for nuclear reactions within stars to produce elements heavier than helium, which form the building blocks of complexity. Because the universe is expanding, it also must be at least billions of light-years in size. We can see how these necessary conditions for the existence of life in the universe also play a role in fashioning the view of the universe that any conscious life-forms will be presented with.

In our own case, the fact that the universe is so big and old has influenced our religious and metaphysical thinking in countless ways. The fact that the universe is big and sparse also ensures that the night sky is dark and patterned with groups of stars, and this feature of our environment has played a significant role in the development of many mythologies and attitudes toward the unknown.[13] It has also played an important role in the course of evolution by natural selection on the Earth's surface. Many of the general ideas that we have about the nature of the universe, and its origins, have their basis in these metaphysical attitudes that have been subtly shaped within our minds by the cosmic environment about us.

Technological Limits

Even if we were able to conceive of and formulate a theory of everything—perhaps as a result of principles of logical self-consistency and completeness alone—we would be faced with an even more formidable task: testing it by experiment. There is no reason why the universe should have been constructed for our convenience. The decisive features predicted by such theories might well lie beyond the reach of our technology. We appreciate that this is no idle speculation.

The United States Congress cancelled the superconductor supercollider project on the grounds of its cost, and the future of experimental particle physics now rests in the hands of the Large Hadron Collider (LHC) project planned for the CERN laboratory in Geneva. These projects seek to increase collider energies far into the tera electron volt range to search for evidence of supersymmetry, the top quark, and the Higgs boson—all pieces in the standard model of elementary particle physics.[14] However, even the energies expected in the LHC fall short by a factor of about one million billion of those required to test directly the pattern of fourfold unification proposed by a theory of everything. Restrictions of economics and engineering, the pressing nature of other more fruitful and vital lines of scientific inquiry, and the collective wishes of the voters in the large democracies doom such direct probes of the ultrahigh energy world of "grand unification."

Unfortunately for us, the most interesting and fundamental aspects of the laws of nature are intricately disguised and hidden by symmetry-breaking processes that occur at energies far in excess of those experienced in our temperate terrestrial environment. The laws of nature are only expected to display their true simplicity at unattainably high energies. According to some superstring theories, the situation is even worse: The laws of nature are predicted only to exhibit that simplicity and symmetry in more space dimensions than the three that we inhabit. The universe may have nine or twenty-five spatial dimensions, but only three of those dimensions are now large and visible; the rest are confined to tiny length-scales, far too small for us to scrutinize directly.[15] Our best hope of an observational probe is if there are processes that leave some small, but measurable, traces that trickle down to very low energies.

Alternatively, particle physicists look increasingly at astronomical environments to produce the extreme conditions needed to manifest the subtle consequences of their theories. In theory, the expanding universe experienced arbitrarily high energies during the first moments of its expansion. The nature of a theory of everything would have influenced the character of those first moments and may have imprinted features upon the universe that are observable today.[16] However, the most promising theory of the behavior of the very early universe—the inflationary universe—dashes such hopes. Cosmological inflation requires the universe to have undergone a brief period of accelerated expansion during its early expansion history.[17] This has the advantage of endowing the universe with various structural features that we can test today. However, it also has the negative effect of erasing all information about the state of the universe before inflation occurred. Inflation wipes the slate clean of all the information the universe carries about the Planck epoch when the theory of everything would have its principal effects upon the structure of the universe.[18]

Limits upon the attainment of high energies are not the only technological restrictions upon the range of experimental science. In astronomy, we appreciate that the bulk of the universe exists in some dark form whose existence is known to us only through its gravitational effects.[19] Some of this material is undoubtedly of very faint objects and dead stars, but the majority is suspected to reside in populations of very weakly interacting particles. Since there are limits on our abilities to detect very faint objects and weakly interacting particles in the midst of other brighter objects, and strongly interacting particles, there may be a technological limit upon the extent to which we can determine the identify of the material content of the universe. Similarly, the search for gravitational waves might, if we are unluckily situated in the universe, require huge laser interferometers or space-based probes in order to be successful and might therefore ultimately be limited by costs and engineering capabilities in just the same way that high-energy physics has proved to be.

At present, we know of four forces of nature and attempts to create unified theories of physics that join them into a single superforce described by a theory of everything that assumes only these four forces exist. But there could exist other forces of nature, so weak that their effects are totally insignificant both locally and astronomically, but whose presence in the theory of everything is crucial for determining its identity. There is no reason why the forces of nature should all have such strengths that we can detect them with our present technology.

It is interesting to recall how accidents of our own location in the universe have made the growth of technical science possible in many areas. Even if some hypothetical extraterrestrial life-forms required high intelligence to survive in their local environment, we should not assume that this means that they will have highly developed scientific knowledge in all areas.

For example, it is an accident of geology that our planet is well endowed with accessible surface metals. Without them, no technology would have developed. The existence of the Earth's magnetic field, together with the presence in the Earth's crust of magnetic and radioactive materials, has led to our understanding of these forces of nature. Similarly, an accident of meteorology has saved the Earth from a permanent sky covering of cloud and enabled us to develop astronomical understanding. And, even with cloud-free skies, we have benefited from an accident of astronomy: Our particular location in the disk of the Milky Way could easily have been shrouded by dust in all directions (rather than just in the plane of the Milky Way), so inhibiting the development of optical astronomy.

Our ability to test Einstein's theory of gravitation hinged originally upon two coincidences about our solar system. The fact that the full Moon and the Sun have the same apparent size in the sky (despite being very different in real size and in distance from us) means that complete eclipses of the Sun can be seen from Earth. This enabled the light-bending predictions of Einstein's general theory of relativity to be tested in the first half of this century. Similarly, the presence of a planet with an orbit as close to the Sun as Mercury's enabled the predictions of perihelion precession by planetary orbits to be checked. Without these fortunate circumstances, scientists would have been left with a largely untestable theory. These examples are given simply to make the point that scientific progress is not necessarily an inevitable march of progress that occurs in any civilization that is "advanced" by some criterion. There may be accidents of environment that prevent the development of science in some directions, while facilitating it in others.

There have been many speculations about the physical nature of extraterrestrial beings and discussions about the likelihood of their existence.[20] There also have been interesting discussions about their ethical imperatives and Christian theologians from St. Augustine to C.S. Lewis have engaged the question of the place of extraterrestrials in their theology. All sorts of answers were arrived at: Augustine used the assumption of the uniqueness of the Incarnation to argue that extraterrestrial life did not exist; the rationalist Thomas Paine claimed that the existence of extraterrestrial life was obvious and hence the Incarnation could not have occurred; C.S. Lewis, in his famous trilogy of science fiction stories, explored the idea that only humans had experienced the Fall and were moral pariahs in a universe of unfallen beings.[21]

Fundamental Limits

One hallmark of the progress of a mature science is that it eventually begins to appreciate its own boundaries. In the present century, we have seen many examples of this. In an attempt to extend a theory in new ways what has often been discovered is some form of "impossibility" theorem—a proof using the assumptions of the theory that certain things cannot be done or certain questions cannot be answered. The most famous examples are Heisenberg's Uncertainty Principle in physics, Einstein's speed-of-light limit on signalling velocities, Gödel's incompleteness theorem in mathematical logic, Arrow's theorem in economics, Turing's uncomputability, the intractability of nondeterministic polynomial (NP) time problems, like the "travelling salesman problem," and Chaitin's theorem about the unprovability of algorithmic randomness. It may transpire that these impossibility results, together with many others that are suggested by studies of space-time singularities, space-time horizons, and information theory, may place real restrictions on our ability to frame or test a theory of everything.

We already know the finite velocity of light ensures that we have a visual horizon (about 15 billion light-years away) in the universe, beyond which light has not had time to reach us since the expansion of the universe began. Thus, we are always prevented from ascertaining the structure of the entire universe (which may be infinite in extent). Astronomers are confined to studying a finite portion of it, called the visible universe. It may be that the visible universe does not contain enough information to characterize the laws of physics completely. Certainly, it does not carry enough information to determine the nature of any initial state for the whole universe without the introduction of unverifiable, and necessarily highly speculative, "principles" to which the initial state is believed to adhere.[22] Many such principles have been proposed—the "no boundary condition" of Hartle and Hawking, the minimum gravitational entropy condition of Penrose, and the outgoing wave condition of Vilenkin are well-known examples.[23] Global principles of this sort all provide quantum-averaged specifications of the entire cosmological initial state. But our visible universe today is the expanded image (possibly reprocessed by inflation) of a tiny part of that initial state, in which conditions may deviate from the average in some way, if only because we know that they happened to satisfy the stringent conditions necessary for the eventual evolution of living complexity.

As yet, Gödelian incompleteness has not placed any restriction upon our physical understanding of the universe, although it has the scope to do so, because it has been shown by Chaitin[24] that it can be recast as a statement that sequences cannot be proved to be random (a Gödel undecidable statement might be just the one needed to characterize the order in a sequence). Gödel's theorem requires that logical systems large enough to contain the whole of arithmetic are either inconsistent or incomplete. Now, although modern physicists use the whole of arithmetic (and much more besides) to describe the physical universe, we cannot conclude from this that there will exist some undecidable proposition about the universe. The description of the laws of nature may require only the decidable part of mathematics.

Alternatively, it may not be necessary to use a mathematical structure as rich as the whole of arithmetic to characterize the universe. The fact that we do may simply be because it is quick and convenient to do so or because of a sequence of historical accidents that have bequeathed a particular mathematical system. If one uses a smaller logical system than arithmetic (like Euclidean geometry or Presburger arithmetic, with only + and - operations), then there is no Gödel incompleteness: In these simpler axiomatic systems, all statements can be demonstrated to be true or false, although the procedure for deciding may be very lengthy and laborious.

Recently, examples have arisen in the study of quantum gravity[25] in which some quantity that is observable, in principle, is predicted to have a value given by a sum of terms that is uncomputable in Turing's sense (the list of terms in the sum could only be provided by a solution of a classic problem that is known to be incomputable —the list of all compact four manifolds). Again, while this may have fundamental significance for our ability to predict, we cannot be sure that it imposes an unavoidable restriction. There may exist another way of calculating the observable in question using only conventional computable operations. Another interesting example arises in the biochemical realm, where it is known that nature has found a way to fold complex proteins quickly. If it carries out this searching process by the process that we would use to program it computationally, then it would seem to be incomputable in the same way that the "travelling salesman problem" is incomputable. However, nature has found a way to carry out the computation in a fraction of a second. As yet we do not know how it is done. Again, we are led to appreciate the difference between limitations upon nature and limitations upon the particular mathematical computational or statistical descriptors that we have chosen to describe its behavior.

Discussion

We have explored some of the ways in which the quest for a theory of everything in the third millennium might find itself confronting impassable barriers. We have seen there are limitations imposed by human intellectual capabilities, as well as by the scope of technology. There is no reason why the most fundamental aspects of the laws of nature should be within the grasp of human minds, which evolved for quite different purposes, or why those laws should have testable consequences at the moderate energies and temperatures that necessarily characterize life-supporting planetary environments.

There are further barriers to the questions we may ask of the universe, and the answers with which it can provide us. These barriers are imposed by the nature of knowledge itself, not by human fallibility or technical limitations. The fundamental limits to human knowledge seem to emerge from mathematics and psychology that appear at first sight to have little in common. But they are the only two subjects that are able to make statements about themselves. The implications of this self-reference problem for mathematics and logic are spread out into other aspects of human thought by Popper and MacKay.[26] MacKay has used them to frame an interesting argument about the impossibilities of certain sorts of predestination and foreknowledge. It is impossible for our future to be constrained by a prediction of its course if we are aware of the prediction; we can always falsify it. However, if that prediction is not made known to us, then it can remain valid.

As we probe deeper into the intertwined logical structures that underwrite the nature of reality, we can expect to find more of these deep results that limit what can be known. Universes that contain complexity and life must possess order of some sort. That order is equivalent to the existence of some natural mathematical structure. If that structure has a complexity above a certain critical level (sufficient to define ordinary arithmetic), then there must be some incompleteness in our ability to determine the truth or falsity of statements about the universe as long as all the truths of mathematics are instantiated in facets of the physical universe.

Notes

1. This story seems to have arisen from a misinterpretation of a letter of resignation written by Commissioner Henry L. Ellsworth to his employers, the U.S. Patent Office. The story is scotched by E. Jeffery in "Nothing Left to Invent," *Journal of the Patent Office*, 22 (1940): 479-481.
2. J.D. Barrow, *Theories of Everything* (Oxford: Oxford University Press, 1991); J.D. Barrow, *The World Within the World* (Oxford: Oxford University Press, 1988).

3. Ibid.
4. J. D. Barrow, *The Artful Universe* (Oxford and New York: Oxford University Press, 1995).
5. M. Green, J. Schwartz, and E. Witten, *Superstrings,* (Cambridge: Cambridge University Press, 1987); F.D. Peat, *Superstrings and the Search for a Theory of Everything* (Chicago: Contemporary Books, 1988); M. Green, "Superstrings," *Scientific American,* September 1986, p. 48; D. Bailin, "Why Superstrings?", *Contemporary Physics,* 30 (1989) 237; P.C. W. Davies and J.R. Brown, eds., *Superstrings: A Theory of Everything?* (Cambridge: Cambridge University Press, 1988).
6. C. Froggatt and H.B. Nielsen, *Chaotic Gauge Theories* (Singapore: World Scientific, 1990); J.D. Barrow, and F.J. Tipler, *The Anthropic Cosmological Principle* (Oxford: Oxford University Press, 1986).
7. C. Isham in R.J. Russell, W. Stoeger, and G.V. Coyne, *Physics, Philosophy and Theology* (Notre Dame, Ind.: University of Notre Dame Press, 1988); S.W. Hawking, *A Brief History of Time* (New York: Bantam, 1988).
8. Green and Schwartz, *Superstrings.*
9. Barrow, *Theories of Everything;* Barrow, *The World Within the World.*
10. Barrow, *The Artful Universe.*
11. J.D. Barrow, *Pi in the Sky* (Oxford and New York: Oxford University Press, 1992)
12. Ibid.
13. Barrow, *The Artful Universe.*
14. L. Lederman, *The God Particle* (New York: Bantam, 1993).
15. See Note 4 on Superstrings; J. D. Barrow, "Observational Limits on the Time-Evolution of Extra Spatial Dimensions," *Physical Review* D 35 (1987): 1805.
16. J. D. Barrow, *The Origin of the Universe* (London: Orion, 1994).
17. A. Guth, "The Inflationary Universe: A Possible Solution to the Horizon and Flatness Problems," *Physical Review* D, 23 (1981):347; J.D. Barrow, "The Inflationary Universe: Modern Developments," *Quarterly Journal of the Royal Astronomical Society,* 29 (1988):101; A. Guth and P. Steinhardt, "The Inflationary Universe," *Scientific American,* May 1984, pp. 116-120; S.K. Blau and A. Guth in S.W. Hawking and W. Israel, *300 Years of Gravitation* (Cambridge: Cambridge University Press, 1987), pp. 524-597.
18. Barrow, *The Origin of the Universe.*
19. L. Krauss, *The Fifth Essence: The Search for Dark Matter in the Universe* (New York: Basic Books, 1989).
20. See the collection of articles *Extraterrestrials: Science and Alien Intelligence,* ed. E. Regis (Cambridge: Cambridge University Press, 1985); P.C.W. Davies, *Are We Alone* (London: Penguin, 1995); J.D. Barrow and F.J. Tipler, *The Anthropic Cosmological Principle.*

21. C.S. Lewis, *Out of the Silent Planet; Perelandra; That Hideous Strength* (London: Macmillan, 1938, 1944, 1946).
22. Barrow, *The Origin of the Universe* and "Unprincipled Cosmology."
23. J. Hartle and S.W. Hawking, "The Wave Function of the Universe," *Physical Review* D, 28, (1983): 2960; S.W. Hawking, *A Brief History of Time* (New York: Bantam, 1988); J. Halliwell, "Quantum Cosmology and the Creation of the Universe," *Scientific American*, December 1991, pp.28-35; R. Penrose, *The Emperor's New Mind* (Oxford: Oxford University Press, 1989); A. Vilenkin, "Boundary Conditions in Quantum Cosmology," *Physical Review* D, 33 (1982): 356.
24. Barrow, *Pi in the Sky*; G. Chaitin, *Algorithmic Information Theory* (Cambridge: Cambridge University Press, 1987; G. Chaitin, "Randomness in Arithmetic," *Scientific American*, July 1980, pp. 80-85; J. Casti, *Searching for Certainty* (New York: Morrow, 1990).
25. J. Hartle and R. Geroch in *The Quantum and the Cosmos - J.A. Wheeler Festschrift*, ed.W. Zurek (Princeton: Princeton University Press, 1988).
26. K. Popper, "Indeterminism in Quantum Physics and in Classical Physics," *British Journal for the Philosophy of Science* 1 (1950): 117 and 1(1950): 173; D. M. MacKay, "Choice in a Mechanistic Universe," *British Journal for the Philosophy of Science*, 22 (1971): 275; D.M. MacKay, "The Logical Indeterminateness of Human Choices," *British Journal for the Philosophy of Science* 24 (1973): 405; D. M. MacKay, *Freedom of Action in a Mechanistic Universe*, Eddington Lecture (Cambridge: Cambridge University Press, 1967); D. M. MacKay, *Brains, Machines and Persons*, chap. 5 (Grand Rapids, Mich.: Eerdmans, 1980); D. M. MacKay, *The Clockwork Image*, chap. 8 (Downers Grove, Ill.: Intervarsity Press, 1974).

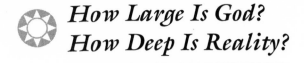

How Large Is God?
How Deep Is Reality?

Robert L. Herrmann

The Impact of an Enormous Universe

Two years ago, Sir John Templeton and I wrote a book about reality, with the intent of showing that two of the most revered sources of truth—religion and science—fell far short of giving us anything even remotely approaching a full understanding of the world and our place in it.[1]

We readily agreed that science had taught us wonderful things, but we also pointed out that it had brought us more questions than answers, more mysteries and less certainty about our place in a universe far larger than we could have imagined. Furthermore, these magnitudes also brought into question the theological presuppositions of our central place in the cosmic scheme of things. Why should the Creator of the universe single out one tiny planet in one obscure little star in one of billions of galaxies as the focal point for the divine revelation? And here, too, our conception of the Creator and our knowledge of the Divine Order must surely be just as minuscule as the scientists' knowledge of the vast reaches of space.

Timothy Ferris, in his book, *Coming of Age in the Milky Way*, talks about our ignorance in light of the enormous size of the universe:

> And yet the more we know about the universe, the more we come to see how little we know. When the cosmos was thought to be but a tidy garden, with the sky its ceiling and the earth its floor and its history coextensive with that of the human family tree, it was still possible to imagine that we might one day comprehend it in both plan and detail. That illusion can no longer be sustained. We might eventually obtain some sort of bedrock un-

217

derstanding of cosmic structure, but we will never understand the universe in detail; it is just too big and varied for that. If we possessed an atlas of our galaxy that devoted but a single page to each star system in the Milky Way (so that the sun and all its planets were crammed on one page), that atlas would run to more than ten million volumes of ten thousand pages each. It would take a library the size of Harvard's to house the atlas, and merely to flip through it, at the rate of a page per second, would require over ten thousand years. Add the details of planetary cartography, potential extraterrestrial biology, the subtleties of the scientific principles involved, and the historical dimensions of change, and it becomes clear that we are never going to learn more than a tiny fraction of the story of our galaxy alone and there are a hundred billion more galaxies. As the physician Lewis Thomas writes, "The greatest of all the accomplishments of twentieth-century science has been the discovery of human ignorance." [2]

A Fundamental Change in Our View of Truth-Seeking

But the recognition of the sheer size of the universe is not the only factor that has brought about an appreciation of the severe limitations of our knowledge. Arthur Peacocke has written in his book, *Intimations of Reality*, that a fundamental change has come about in the past twenty-five years in our view of reality in both science and theology.[3]

In the sciences, it had been naively assumed that truth could be perceived directly by means of theory and observation, even if the observations were indirect. Peacocke quotes philosopher Mary Hesse's description of the earlier view as given in her Revelations and Reconstructions in the Philosophy of Science:

> Science is ideally a linguistic system in which true propositions are in one-to-one relation to facts, including facts that are not directly observed because they involved hidden entities or properties, or past events or far distant events. These hidden events are described in theories, and theories can be inferred from observa-

tion, that is, the hidden explanatory mechanism of the world can be discovered from what is open to observations. Man as scientist is regarded as standing apart from the world and able to experiment and theorize about it objectively and dispassionately. [4]

Hesse goes on to explain that the past two decades have seen momentous developments that made this definition appear exceedingly naive. Thomas Kuhn published his *Structure of Scientific Revolutions* in 1970, in which he proposed a new interpretation of the history of science.[5] He argues that science goes through periods of normality during which accepted paradigms—broad conceptual frameworks—are employed and applied, and periods of rapid change, or revolutions, in which these paradigms are shattered and replaced by new ones.

Following Kuhn, a new emphasis was placed on sociological factors that strongly influenced the development of science. Peacocke describes it as follows:

> Science came to be seen as a continuous social enterprise, and the rise and fall of theories and the use and replacement of concepts as involving a complex of personal, social, intellectual, and cultural interactions that often determined whether a theory was accepted or rejected. Theories are constructed, it was argued, in terms of the prevailing "world view" of the scientist involved: so to understand them one must understand the relevant world view. A new emphasis was therefore placed on the history of science, especially the sociological factors influencing its development. Thus a new area was opened up for the application of the expanding enterprise of the sociology of knowledge in general and of scientific knowledge in particular. However, it turns out that the "world view" of the scientist is an exceedingly complex and elusive entity even more so when a community of scientists is involved. [6]

Some sociologists of science even proposed that the products of science are purely social constructions and have little bearing on the real nature of the physical world.

A Unique Role for Science

Against this extreme view of knowledge, which of course would be true of any area of exploration (perhaps also of the historian of science!) other philosophers of science such as Ernan McMullin have argued for a unique position for scientific truth. McMullin argues that although science is a social product, it has the characteristic of self-correction, a continuous filtering and sifting that goes on in the course of experimental collaboration and publication.[7] Then, too, scientific theories are valued on the basis of their "fertility," their ability to predict new experimental directions and to suggest modifications to existing theories.

In this latter aspect, a theory is not unlike the poet's metaphor, which the poet uses to help illuminate something not well understood by providing a special kind of analogy. The manner in which such metaphors work is by tentative suggestion, engaging both poet and reader in a creative thinking process. The science writer K.C. Cole quotes Niels Bohr as once having said, "When it comes to atoms, language can be used only as in poetry. The poet too is not nearly so concerned with describing facts as with creating images."[8]

That something rather special is going on in scientific truth-seeking is further hinted at by scientist-philosopher Michael Polanyi in his book, *Personal Knowledge*. Polanyi talks about a faculty that is required to arrive at truth in any field. That faculty is faith—a commitment to finding the truth—a feeling after the truth by way of a "passionate contribution in the personal act of knowing."[9] What Polanyi noted as a common feature of all theorizing in science was belief in and commitment to scientific theories as potentially true; this has always been a critical aspect of scientific discovery. But the most telling feature of the idea of personal knowledge is revealed by Polanyi's analysis of what he calls tacit knowledge. He notes that underneath the judgmental and perceptual skills that are applied by the scientist are a set of inarticulate skills and arts that are essential to our theorizing.

Criteria such as symmetry, simplicity, elegance, fruitfulness, and satisfaction are not susceptible to logical scrutiny, but they form a significant component of our theorizing. These tacit components may be viewed in the aggregate as different aspects of what the scientist often terms beauty, the sense of which we are often unaware as we seek to build a theoretical framework for our observations. In the words of Walter Thorson:

> Our sense, and the collective tradition, of beauty and hence the character of our tacit criteria is capable of change and development; but unmistakably it is a sense of beauty which moves us to prefer some theories to others, and even to heuristically commit ourselves to them, even though as yet we have no clear conception of their consequences. Now it is a surprising thing that this general expectation regarding reality is not disappointed far more often than it is rewarded, but on the contrary it seems to have a real power to evoke creative vision within the human mind.[10]

The notion that we can be led in the direction of reality by such rather nonobjective criteria may seem mysterious, but it also does seem to provide another argument for the idea that scientific truth-seeking is much more than a social construct.

The Critical Realist Position

The debates between extremists arguing for scientific reality as purely social construction and those arguing for science as final truth have led to a generally accepted view referred to as the critical realist position. McMullin describes this as follows:

The basic claim made by scientific realism . . . is that the long-term success of a scientific theory gives reason to believe that something like the entities and structure postulated in the theory actually exist. There are four important qualifications built into this: (1) the theory must be a successful one over a significant period; (2) the explanatory success of the theory gives some reason, though not a conclusive warrant, to believe [it]; (3) what is believed is that the theoretical structures are something like the structures of the real; (4) no claim is made for a special, more basic, privileged form of existence for the postulated entities.[11]

According to Peacocke, the qualifications ("significant period," "some kind," "something like") built into this description are essential to a defense of scientific realism. Working scientists understand the extent and nature of provisionality that is assigned to theories in their field. Their proposed theories and models are understood as no more than approximations of the structures of reality. Nevertheless, there is often an increasing sense of confidence that reality is being more and more accurately described, especially as that knowledge is successfully applied in prediction and control and in the design of new experiments. This description Peacocke refers to as "a skeptical and qualified realism."

To illustrate this approach to scientific realism, Peacocke takes the example of the electron. Scientists are committed to "believing in" the electron on the basis of much experimental data. However, "what they believe" about electrons has undergone many changes, but they still refer by way of long social links to the entity described in that historical event when the electron was first discovered. Physicists believe in the existence of electrons but are reluctant to say what they are. But "they are always open to new ways of thinking about them that will enhance the reliability of future predictions and render their understanding more comprehensive."[12]

But what makes the electron most believable by the physicist is that it functions as a tool. Despite the fact that the electron cannot be observed, it can regularly be manipulated to investigate other aspects of nature. Peacocke calls this "the experimental argument for realism," pointing out that it deals with entities rather than theories.[13] To illustrate the experimental argument, Peacocke quotes Ian Hacking in *Representing and Intervening* as follows:

> There are surely innumerable entities and processes that humans will never know about. Perhaps there are many that in principle we can never know about. Reality is bigger than us. The best kinds of evidence for the reality of a postulated or inferred entity is that we can begin to measure it or otherwise understand its causal powers. The best evidence, in turn, that we have this kind of understanding is that we can set out, from scratch, to build

machines that will work fairly reliably, taking advantage of this or that causal nexus. Hence, engineering, not theorizing, is the best proof of scientific realism about entities.[14]

Scientific Models

We have alluded to the way in which scientific theorizing and model building had its counterpart in the poet's metaphor, which helps to illuminate that which is not well understood by tentative suggestion, leaving the reader to create ideas and images from what is given in another more familiar form. The scientific model performs a similar function in that it generates metaphorically theoretical terms that suggest a network of explanation. Scientific theories are built by construction of models, and good models are not only inseparable from their theories but also allow for theory development by suggesting new possibilities and accommodating new observations. "A model," Janet Soskice says in *Metaphor and Religious Language*, "is the living part of the theory, the cutting edge of its projective capacity and, hence,. . . indispensable for explanatory and predictive purposes."[15]

Peacocke points out that there is no total agreement on the essential role of the model in scientific theorizing. Some scientists and philosophers regard models as helpful but carrying out no commitment concerning their relation to reality. But the majority take a critical realist approach and see models as essential and permanent features of science. However, he points out that the critical realist sees the model as only a partial and incomplete expression of reality, a construction that no one would view as final truth. The use of complementary models in science, such as the wave and particle models for light in physics, should serve to remind us that models can never be literal and always contain some inadequacy.[16]

In agreement with Peacocke, physicist Richard Morris, in his book *The Nature of Reality*, points out that by the end of the nineteenth century, physicists were viewing scientific theories as approximations that were only models of reality, although sometimes very good ones.

He comments:

> Theories bore the same relation to the physical world that an architect's model did to a building, or that a detailed drawing did to a scene in nature. The correspondence between model and reality might be very close, but it could never be exact.[17]

Science as a Hierarchy of Systems

Science has been a highly successful enterprise in part because one of the chief methodologies involves the resolution of complex structures into simpler components. However, the success of this reductionism at the methodological level often carries over into a more general philosophical attitude. In the case of biology, for example, one finds biologists saying that living organisms are "nothing but" a physio-chemical system, and some sociobiologists saying that all social behavior is explainable on a biological basis. Against this attitude, Peacocke argues for an appreciation of the distinctiveness of the various scientific disciplines as an ascending arrangement or hierarchy of systems.

In the progression from physics to chemistry to biology to ecology, each succeeding science becomes enriched in empirical content and new concepts emerge at each level that simply are not present in the preceding levels. Statements that are true of the lower levels are also true of the higher ones, although the lower level statements are usually not the focus of interest for scientists working in the higher level disciplines. However, concepts appear at the higher level that have no definition at the lower. For example, British biophysicist Donald MacKay points out, in *The Clockwork Image,* that a term from biology like adaptation has no definition in physics and chemistry.[18] Likewise, Peacocke uses biological information as illustrative: Our knowledge of the physiochemical nature of genetic material in the form of deoxyribonucleic acid is a salient contribution from chemistry, but the concept of information flow is a function and property of the whole organism and is necessarily restricted to the level of biology or above. We can be reductionists in methodology but not in epistemology. Theories are autonomous.[19]

One interesting consequence of the existence of hierarchies of complexity in science is that we can see the natural world as an evolved series of hierarchies in which new kinds of complex organization arise from simpler systems over vast periods of time and by distinct developmental steps. A second consequence of hierarchical structure is its significance for our ideas about scientific reality. The enrichment of our knowledge of reality with each succeeding order of complexity, bringing with it new concepts and methodologies, greatly broadens our view of reality. The reductionist view that the only "real" explanations are at the level of atoms and molecules would appear to leave us grossly impoverished.

The Elusiveness of Final Answers in Current Science
Particle Physics Displays Multiplied Complexity

Quarks

Since the late 1940s, the number of new kinds of particles has multiplied incredibly. Further studies of cosmic ray showers and experiments with cyclotrons yielded a variety of mesons and a large number of heavier particles that were similar to the proton and neutron. The latter are now referred to as baryons. By the end of the 1950s, so many particles had been discovered that the term elementary particle was clearly inappropriate. There were hundreds of different mesons and hundreds of different baryons. And there was no visible limit to the number of particles that might be found in the future. Robert Crease and Charles Mann tell us that by 1958 there were so many new particles that scientists at the University of California at Berkeley began to publish an almanac. The first edition was nineteen pages long and discussed sixteen new particles. The edition of 1984 ran three hundred and four pages and referred to two hundred particles.[20]

A major step in the direction of simplification was desperately needed in the physics of quantum particles, and it was partially achieved by the classification of particles according to the forces that they experienced, particularly the strong and weak nuclear forces. Particles that were affected by the strong force were called hadrons, and included both baryons—heavy particles like the proton and neutron—

and mesons. Particles that felt only the weak force were called leptons. The lepton family was made up of the electron, the muon (a particle found in cosmic rays and originally confused with the pion) and the neutrino.

This classification brought some relief to particle physicists, although it still left the hadron category far too large. A second step for subdivision of the hadron group was proposed in 1961 by American physicist Murray Gell-Mann and Israeli physicist Yuval Ne'eman. It allowed for the most commonly observed baryons and mesons to be put together in sets of eight and was called, somewhat facetiously, the eightfold way. The original eightfold way was the a path to enlightenment devised by the Buddha in the fifth or sixth century BC. The proposal was made that hadrons were separable into groups by virtue of the presence of different subatomic constituents. Gell-Mann called these components quarks. Originally there were three of these: up, down, and strange, and each had an antiparticle.

According to the theory, which was elaborated by Gell-Mann and independently by American physicist George Zwieg, all the hundreds of baryons and mesons could be explained as combinations of six types of quarks and antiquarks. Baryons were made up of three quarks, mesons of a quark and an antiquark. The proton, for example, contained one down and two up quarks, while the positively charged pion would have an up and an antidown quark. The large number of particles in the baryon and meson categories can be explained by this scheme with the added qualification that energy changes lead to slightly different particles. Many of the new particles were nothing more than higher energy states of the old particles.

The existence of quarks was verified in 1968 in an experiment performed with the Stanford Linear Accelerator. However, this relatively simplified quark theory was short-lived. From 1974 to 1984, three more quarks were found to be necessary to explain newly discovered hadrons, and, in addition, all six quarks appeared to carry an additional characteristic called color. They could be either red or green or blue. The color designations did not imply actual color. Quarks are too small to be involved in light interactions but were simply three additional charge-related properties that were part of the character of quarks. The upshot of this refinement was that there were now eighteen different quarks. Quarks came in six flavors, as they were called: up, down, strange, charmed, bottom, and top; and three different colors: red, green, and blue.[21]

Virtual Particles

The attempt to classify quantum particles according to the forces they experienced led to an additional powerful insight about the relationships between subatomic particles. As quantum field theorists examined the behavior of the four known fields of force—electromagnetism, the strong nuclear force (which holds the nucleus together), the weak nuclear force (involved in radioactive disintegration), and gravity—they began to realize the full implications of Werner Heisenberg's Uncertainty Principle. Not only is the principle applicable to the velocity and momentum of an electron but also to any pair of what are called conjugate variables; any pair of variables that are related in a certain reciprocal way. Energy and time are conjugate variables, which means that the energy of a particle cannot be precisely defined over extremely short time periods. In fact, if the time span is short enough, the uncertainty in the energy may be great enough to create new particles out of nothing. These particles are called virtual particles, and their existence is transient indeed. As Morris describes them: "They spring into existence for tiny fractions of a second, deriving the energy required for their existence from Heisenberg uncertainties. Naturally, they cannot exist for long, for this energy debt soon has to be paid back. When it is, the virtual particles disappear into nothingness again." [22]

It should be stressed that the virtual particle is not viewed as any less real during its brief existence than electrons or protons or neutrons, although the latter are often referred to as real particles. Quantum field theorists explain that virtual particles are constantly coming into existence everywhere, even in the empty space of a perfect vacuum. Thus, in quantum theory, the term nothing has no meaning. These implications of quantum theory appalled many scientists. As Crease and Mann describe it:

> The Uncertainty Principle exposed a frightful chaos in the lowest order of matter. The spaces around and within atoms, previously

thought to be empty, were now supposed to be filled with a boiling soup of ghostly particles. Small wonder that many theorists, including Einstein, were appalled. From the perspective of quantum field theory, the vacuum contains random eddies in the field of space-time: tidal whirlpools that occasionally hurl up bits of matter only to suck them down again. Since these bits virtually don't exist, they have been named "virtual particles."[23]

But these authors go on to point out that virtual particles are far from a bizarre anomaly. In fact, they are a key feature of quantum field theory. Heisenberg, Wolfgang Pauli, and others saw that electromagnetism, the force responsible for light, electricity, and magnetism, can be mediated by the transfer of photons between charged particles. More recently, quantum field theorists have elaborated this concept in a theory called "quantum electrodynamics." The theory proposes that electrons can spontaneously emit virtual photons, and these photons can be absorbed by other charged particles.

The way in which these forces act may be illustrated by means of the analogy of two skaters throwing a ball back and forth. As the ball is exchanged, each skater is forced backward, as though a force of repulsion existed between them. Alternatively, if the skaters turned their backs to one another, and threw a boomerang back and forth, the changing direction of the boomerang would force them toward each other. This would be analogous to the force of attraction between a negatively and a positively charged particle. Like all analogies, this one fails in that it implies too much about our knowledge of the path followed by the photon, since this is part of the uncertainty that is always present with quantum particles.

Further Complexities from the Quantum Electrodynamics Theory

The major uncertainty of the quantum electrodynamics formulation stems from the fact that the theory predicts certain infinities. For example, when the energy of interaction between an electron and the cloud of virtual particles surrounding it is computed, the result is infinity. In addition, the energy of "empty" space is predicted to be infinite. That is, the energy associated with the appearance of a virtual particle, in the absence of matter, is infinite. Theorists generally see such infinities as signals that something is wrong. However, German physicist Victor Weisskopf suggested a mathematical trick called renormalization, in which certain infinite quantities were subtracted from each other to give finite results. Many scientists objected, including Paul Dirac, who pointed out that a sensible mathematical approach is to neglect a quantity when it is small, but not when it is infinite. Critics also point out that if we take the results of renormalized quantum electrodynamics (QED) literally, it fails to give us an understandable picture of the nature of subatomic reality.

For example, the properties of the electron appear to include infinite mass, infinite charge, and infinite energy. It has been suggested that one possible reason these properties are not actually seen is that the electron is screened by a cloud of virtual particles. In this way the infinite energy of the "bare" electron may be canceled out by the infinite energy of the virtual particle cloud. It may be that QED is telling us that an electron cannot be defined apart from the field that surrounds it.

A New Kind of Complexity

Efforts at simplification in particle physics by examining the fields of force in which various subatomic particles are involved has led to the concept that the force of electromagnetism is mediated by an interchange of photons between electrons and other charged particles. More recently, a similar situation has been postulated for the strong force that binds protons and neutrons together in the nucleus. Here, interchange of heavy particles called pi-mesons occurs. These interchanges also involve virtual particles.

Finally, the discovery of quarks as subunits of the heavy particles—protons, neutrons, and mesons—has brought a certain simplification but also a new kind of complexity to the picture in particle physics. More recently, a theory analogous to quantum electrodynamics has been proposed to explain how quarks are bound together in the nucleus. In this theory, called quantum chromodynamics (QCD), there are eight different entities called "gluons" that mediate the interaction between quarks in the same manner that photons function as virtual particles in quantum electrodynamics. The situation as described by Morris:

> QCD is quite an elegant theory, but it does not have the simplicity that the physicists who proposed the original version of the quark theory were seeking. Three quarks have not proved to be adequate to explain observed phenomena. Six different quark flavors are now known. According to QCD, each of these six quarks can have any of three different colors. Thus the number of different kinds of quark has been multiplied to eighteen (or thirty-six, if antiquarks are counted separately). In addition, there are eight gluons, making twenty-six particles in all. . . It appears that the search for simplicity has only led to a new kind of complexity. Perhaps quarks are not really the fundamental constituents of matter. There is no reason why quarks could not be made of yet smaller particles. If they are, the same might be true of leptons.[24]

Morris goes on to say that quarks and leptons may be the ultimate constituents of matter. On the other hand, there may be additional levels, and even the possibility that there are an infinite number of levels. Quarks and leptons may be composed of smaller particles, which are composed of yet smaller particles, and so on. Morris concludes, "Physical reality could conceivably be a kind of cosmic onion, with level below level below level. It may be that the levels never come to an end."[25]

Reality as Unity of the Forces of Nature

The Idea of Unification

In *The Critique of Pure Reason,* philosopher Immanuel Kant argued that all human wisdom is limited because our minds must impose a structure on our experience. The consequence of this, in the words of Crease and Mann, is:

> As consequence, we are led inevitably to make certain supposi-
> tions about nature that are, in actuality, by-products of the orga-
> nizing activity of our own brains. Such presuppositions—Kant
> showed their number, and said they apply to more than science—
> are unprovable in theory but indispensable in practice. He called
> them "regulative ideas," and listed the unity of nature as a cardi-
> nal example. Because of it, the structure of science itself draws
> physicists toward unification.[26]

Physicist Albert Einstein had expressed a deep passion to know the mind of God, to know nature as to intent and purpose. Indeed, the search for a unified field theory occupied the greatest part of his later scientific life. For Einstein, there was a remarkable harmony between the human mind and the rationality embedded in the universe, and he sought the explanation for that harmony in the unification of the four forces of nature the strong and the weak nuclear forces, the electro-magnetic force, and gravity.

The beginnings of unification go back to the Scottish physicist and mathematician James Clerk Maxwell, who built a theory of the elec-tromagnetic field on the experimental work of English scientist Michael Faraday. Ferris, in his *Coming of Age in the Milky Way*, tells us that the Edison-like Faraday was a prodigious experimenter, who in the course of forty-six years of research on electricity and magnetism at the prestigious Royal Institution of Great Britain performed more than fifteen thousand experiments.[27] His great contribution was to bring about a fundamental shift in emphasis from the apparatus of electricity, the magnets and coils, to the invisible fields of force that surrounded them.

However, it was Maxwell who provided the deep insights and mathematical skills necessary to build a quantitative structure and to show that electricity and magnetism were aspects of a single force, electromagnetism, and that even light was an expression of this force. This field theory was so unique, so counter to the current preference for mechanical models, that Maxwell felt obliged to invent a mechanical description just for Faraday's benefit. And there was another doubter of equal prestige, Sir William Thompson (Lord Kelvin) who found Maxwell's equations so disturbing that he stated that in departing from mechanical models Maxwell had lapsed into mysticism.

This was an interesting charge, since Maxwell was a fervent evangelical Christian who claimed that he approached his science with the openness of one who expected that God's creation would be rightly studied with an all-important metaphysical reference to the vastness of nature and narrowness of our symbolic sciences. As Thomas Torrance describes it:

> No human science, he felt, could ever really match up in its theoretical connections to the real modes of connection existing in nature, for valid as they may be in mathematical and symbolic systems they were true only up to a point and could only be accepted by men of science, as well as by men of faith, in so far as they were allowed to point human scientific inquiry beyond its own limits to that hidden region where thought weds fact, and where the mental operation of the mathematician and the physical action of nature are seen in their true relation. That is to say, as Clerk Maxwell himself understood it, physical science cannot be rightly pursued without taking into account an all-important metaphysical reference to the ultimate ground of nature's origin in the Creator. Thus while Clerk Maxwell never intruded his theological and deeply evangelical convictions into his physical and theoretical science, he clearly allowed his Christian belief in God the Creator and Sustainer of the universe to exercise some regulative control in his judgment as to the appropriateness and tenability of his scientific theories, that is, as to whether they measured up as far as possible to "the riches of creation."[28]

Could we have, in Maxwell's attitude, another of Kant's regulative ideas? Could the "riches of creation" be another presupposition like the unity of nature? And was it perhaps the more indispensable of the two for James Maxwell? The same question could be asked for Einstein, another deeply religious scientist, who almost single-handedly brought about a revolution in physics in this century. In the theory of special relativity, he showed that mass and energy were equivalent, and in the theory of general relativity he showed that space and time were one inseparable space-time continuum. But he was not successful in his quest for a unified theory.

The Search for Unification

By the last quarter of this century, our understanding of the array of particles and forces and their relationships came to be known as the standard model. The model includes two general categories of particles: fermions, which include protons, neutrons and electrons; and bosons, which convey force and function as exchange particles between fermions.

The standard model is completed by the inclusion of the four fundamental forces—electromagnetism, gravitation, and the strong and weak nuclear forces. The boson for electromagnetism is the photon; that for the weak forces in called a weak boson; that for the strong force is called a gluon, and for gravitation, the graviton.

In principle, every fundamental event in the universe is explained by the standard model. Yet as Ferris tells us in *Coming of Age in the Milky Way,* it is hardly considered the last word by physicists. Ferris quotes physicist Leon Lederman, director of the Fermilab particle accelerator, in a 1965 interview:

> The trouble we're in now is that the standard model is very elegant, it's very powerful, it explains so much but it's not complete. It has some flaws, and one of its greatest flaws is aesthetic. It's too complicated. It has too many arbitrary parameters. We don't really see the creator twiddling seventeen knobs to set seventeen parameters to create the universe as we know it. The picture is not beautiful, and that drive for beauty and simplicity and symmetry has been an unfailing guidepost to how to go in physics.[29]

Then Ferris concludes:

> So it was that physicists late in the century were still searching for a simpler and more efficient account of the fundamental interactions. The object of their quest went by the name "unified" the-

ory, by which they usually meant a single theory that would account for two or more of the forces currently handled by separate
theories. They were guided, to be sure, by experimental data and
by the challenges immediately at hand—the theorist resembles,
as Einstein said, an "unscrupulous opportunist" more often trying to find a specific solution to an immediate problem than to
write a grand explication of everything. But they were guided as
well, as Lederman mentions, by the hope that their accounts of
nature could more nearly approach the elegant simplicity and su-
perlative creativity of nature herself.[30]

Symmetry and Asymmetry

Given the limitations mentioned by Lederman and the general problems with particle physics discussed earlier, physicists clearly were in
need of some more fundamental understanding if a unified theory was
ever to be realized. At this juncture, the most promising clues were
symmetry and asymmetry. In science, the definition for symmetry
could simply be: symmetry exists when a measurable quantity remains
invariant or unchanged under a transformation or alteration. In fact,
all the laws of nature are expressions of invariances and may therefore
involve symmetries.

When the principle was applied to the quantum physics of particles
and fields, it opened up a promising approach to unified field theory.
The key concept was that if any physical system is to demonstrate a
symmetry or invariance, there must exist one aspect that precisely
compensates for changes in another aspect to preserve the invariance.
The logical explanation for the compensation is the action of a force,
presumably mediated by force particles.

But it was soon discovered that there also existed asymmetries in
nature. One of these was discovered by physicists Yang Chea Ning and
Lee Tsung-Dao, who identified a discrete asymmetry in the weak
force that was termed parity violation. Particles emitted by beta decay
spin in one direction. Among those investigating asymmetries was
Steven Weinberg, who along with Abdus Salam and Sheldon Glashow

made the next step in the approach to unified field theory. Each had struggled for a common denominator between two of the four forces, electromagnetism and the weak nuclear force.

Glashow had published in 1961 about the remarkable similarities between the two and predicted the existence of what were called W and Z force-carrying particles, but he was unable to predict their masses. Then it was demonstrated by several workers that symmetry-breaking events could create new kinds of force-carrying particles, some of which could be massive. It was then reasoned that if the particles that carry the weak nuclear and electromagnetic forces were related by a broken symmetry, it must be possible to determine experimentally the masses of the W and Z particles characteristic of the unified, symmetrical force from which the two forces may have arisen. Weinberg had been using the symmetry-breaking ideas, but applying them to the strong force. All at once, in 1967, he realized that the particle descriptions coming from his equations—one set massless, the other massive—did not resemble strong force, but fit perfectly the expectations of the electromagnetic and the weak forces. The massless particle was the photon of the electromagnetic force.

The test of the theory required a particle accelerator of tremendous power, and the two leading candidates were at CERN, the European Center for Nuclear Research, near Geneva, and Fermilab, located in a western suburb of Chicago. It was a CERN accelerator group that eventually found the W and Z bosons. Its leader, Carlo Rubbia, announced in early 1963 that the W particle had been found and the electroweak theory thus confirmed. Shortly thereafter, they found the predicted Z boson as well. Thus, Weinberg, Salam, and Glashow had been proven correct. We do indeed appear to live in a universe of broken symmetries, where at least two of the fundamental forces, electromagnetism and the weak nuclear force, have diverged from a single, more symmetrical parent.

Stephen Hawking, in *A Brief History of Time,* provides another useful perspective on the Weinberg-Salam theory. What appear to be several completely different force particles at low energies are actually all the same type of particle, but in different states. At high energies, all the particles behave the same. At some intermediate energy, what he calls "spontaneous symmetry breaking" occurs. And the particles of the electroweak force differentiate into those of the individual electromagnetic and weak nuclear forces. To illustrate, Hawking provides an analogy from a gambling casino:

> The effect is rather like the behavior of a roulette ball on a roulette wheel. At high energies (when the wheel is spun quickly) the ball behaves in essentially only one way: it rolls round and round. But as the wheel slows, the energy of the ball decreases, and eventually the ball drops into one of the thirty-seven slots in the wheel. In other words, at low energies there are thirty-seven different states in which the ball can exist. If, for some reason, we could only observe the ball at low energies, we would then think that there were thirty-seven different types of ball![31]

Still to be explained is how symmetry breaking leads to the acquisition of mass by the W and Z bosons, although Weinberg has postulated a new particle, the Higgs boson, as the symmetry breaking agent.

Further unification has been attempted by theorists who have sought to join the electroweak theory with the strong nuclear force in a series of proposals referred to as the grand unified theories (GUTs). The first of these was published by Glashow and his American physicist colleague Howard Georgi. As Hawking points out, the title Grand Unified Theory was a bit of an exaggeration. "The resultant theories are not all that grand, nor are they fully unified, as they do not include gravity."[32] The basic concept of the GUTs is that the strong nuclear force gets weaker at high energies whereas the electromagnetic and weak nuclear forces get stronger at high energies. Therefore, at some very high energy, called the grand unification energy, these three forces would all have the same strengths and so could be just different aspects of a single force. The grand unification energy would have to be much more than a million times larger than any particle accelerator can achieve. In fact, the accelerator would need to be the size of the solar system! It is therefore impossible to test grand unified theories directly in the laboratory. Thus far, indirect tests have been unsuccessful.

Critics of the GTUs, like American physicist Julian Schwinger, Glashow's mentor, caution that the unification theory has gone far beyond experiment. He says:

> Unification is the ultimate goal of science, it's true. But that it should be unification now—surely that's the original definition of hubris. There are a heck of a lot of energies we haven't gotten yet. Grand unified schemes make the implicit assumption that what happens in them is not fundamentally different from what we know—an assumption about how nature works that's in conflict with what we've spent the last century learning.[33]

In fact, the GUTs have not proved successful in linking the electroweak and strong nuclear forces, but, as Ferris expresses it, their "waning...went widely unmourned."[34] They have lacked the "sweeping simplicity" that unified theories are supposed to possess. Like the standard model, they were replete with arbitrary parameters, and they omitted the force of gravity. As Freeman Dyson sums it up, "The ground of physics is littered with the corpses of unified theories."[35]

The joining of the fields of particle physics and cosmology brought with it very powerful insights. It was realized that every subatomic particle may have arisen at some point in a process of cosmic evolution. The perplexing array of particles simply attested to the richness of cosmic history. Ferris points out that it is possible to see a direct relationship between the size, binding energy, and age of the fundamental structures of nature.[36] A molecule is larger and more fragile than an atom; atoms are larger and more fragile than nuclei; nuclei are larger but less stable than the quarks of which they are composed.

Cosmology tells us that this is a historical relationship. Quarks were first to be bound together in the extremely energetic early moments of the big bang, followed by the union of protons and neutrons to form the nuclei of atoms as expansion and cooling of the universe occurred, and then electrons found their way into relationship with the nuclei to form atoms that in turn joined together to form molecules. The more closely we examine the structures of nature, the farther back we are looking in time.

It should be noted, too, that as we have moved back in time we also have moved to higher and higher binding energies. Only a few thousand electron volts of energy are required to strip an atom of its electrons. Several million electron volts are required to split up the nucleons, the protons and neutrons of the nucleus, but hundreds of times that energy would be required to dislodge the quarks. The fundamental principle is that the smaller, more fundamental structures are bound by the higher energy levels because they were the structures forged in the heat of the big bang.

The beginning of the universe has been approached scientifically by two related hypotheses: quantum genesis and vacuum genesis. The pioneer of quantum genesis is Stephen Hawking, who holds Newton's chair as Lucasian Professor of Mathematics at Cambridge University.

It was Hawking, together with British physicist Robert Penrose, who played a large part in the formulation of the essential features of the big bang theory of the origin of the universe. Between 1965 and 1970, they focused their attention on the theory of general relativity and on black holes, the ultimate fate of massive stars that had exhausted their fuel. In such a star, the gravitational field at its surface becomes progressively intensified as contraction proceeds, and light rays are finally bent so strongly that they cannot escape. Since relativity specifies that nothing can exceed the speed of light, nothing can escape the gravitational field, and we have what must be a singularity of infinite density and space-time curvature. Hawking pointed out that this was "rather like the big bang at the beginning of time."[37]

However, Hawking has since changed his mind. In particular, he has proposed that what occurred in the first moments of the origin of the universe was not a singularity, a once-and-for- all, unique event, but instead a quantum fluctuation.[38] The impetus for this reexamination comes in part from the frustration that cosmologists feel in seeking to understand physical processes as a singularity. Ordinary physics just does not work at the extremes of energy and density and subatomic dimensions that occur. At this point, the force of gravity, which can be ignored in considering many physical processes, becomes the major factor.

In most of the discussion of unification of the forces, gravity was ignored because it is ordinarily such a weak force in the context of particle-field interactions. However, at the high densities at the start of the big bang, gravity was the major force, just as it is at the collapse of a star to form a black hole. Not only was the force of gravity an important component of the first moments of the big bang but the quantum dimension was considered a crucial factor as well. The problems with bringing in quantum theory is that gravity and quantum behavior are described by Einstein's general theory of relativity and quantum mechanics respectively, two theories inconsistent with each other. For this reason, Hawking proposed a new quantum theory of gravity. This theory makes it possible that the ordinary laws of physics hold everywhere, including the beginning of time. There may be no necessity to postulate new laws for singularities, because as Hawking explains it, there need be no singularities in the quantum approach.[39]

Hawking's proposal was first presented in a 1983 lecture he gave in Padua, where Galileo used to lecture. Ferris describes the theory:

> What emerged was a tale of cosmic evolution possessed of a strangely alien beauty. All world lines diverge from the singularity of genesis, Hawking noted, like longitude lines proceeding from the north pole on a globe of the earth. As we travel along our world line we see the other lines moving away from us, as would an explorer sailing south along a given longitude: this is the expansion of the universe. Billions of years hence the expansion will halt and the universe will collapse, eventually to meld into another fireball at the end of time. There is however, no meaning to the question of when time began or when it will end.[40]

In commenting on the globe analogy at another time, Hawking said that if the suggestion that space-time is finite but unbounded is correct, then the big bang is like the North Pole of the Earth. If we ask what happens before the big bang, it is as meaningless as asking what happens on the surface of the Earth one mile north of the North Pole.

Hawking points out that there is not yet a complete and consistent theory that combines quantum mechanics and gravity.[41] However, one of its features would include physicist Richard Feynman's "sum of histories" approach to quantum mechanics. A particle does not have just a single history in quantum mechanics, as it does in classical theory. Instead, it is supposed to follow every possible path of space-time. So by calculating all possible trajectories of a particle, we arrive at the most probable description of how it reached its present observed state. However, when these calculations are made, serious technical difficulties arise. Their solution is to sum up only those histories that take place in what is called imaginary time.

Actually, Hawking assures us that imaginary time is a well-defined mathematical concept. There are special numbers that give negative numbers when multiplied by themselves (one is called i and when multiplied by itself gives -1, 2i multiplied by itself -4 and so on). For the sum of histories one must, for the purposes of calculation, measure time using imaginary numbers instead of real ones. The curious result is that the distinction between time and space disappears.

A second feature of Hawking's quantum theory of gravity is the use of Einstein's idea of the gravitational field as represented by curved space-time. When Richard Feynman's sum of histories is applied, the analog of the history of a particle is now a complete curved space-time that represents the history of the universe.

Hawking tells us that the end result of combining these features—a sum of histories in imaginary time and curved space-time—is to add one more possibility to the ways the universe can behave. In addition to an infinite existence or to its beginning in a singularity at some time in the past, it is possible in a quantum theory of gravity, with the time directions on the same footing as directions in space, for space-time to be finite in extent and yet have no singularities that formed a boundary. Going back to the global analogy, Hawking says "even though the universe would have zero size at the North and South Poles, these points would not be singularities any more than the North and South Poles of the earth are singular. The laws of science will hold at them , just as they do at the North and South Poles on the earth."[42]

The question of the distinction between real and imaginary time remains. In what sense is the "mathematical device" of imaginary time valid? Hawking suggests that it may even be more basic than real time:

> This might suggest that the so-called imaginary time is really the real time, and that what we call real time is just a figment of our

imaginations. In real time, the universe has a beginning and an end at singularities that form a boundary to space-time and at which the laws of science break down. But in imaginary time, there are no singularities or boundaries. So maybe what we call imaginary time is really more basic, and what we call real is just an idea that we invent to help us describe what we think the universe is like. . . But a scientific theory is just a mathematical model we make to describe our observations: it exists only in our minds. So it is meaningless to ask: Which is real, "real" or "imaginary" time? It is simply a matter of which is the more useful description.[43]

Ferris gives us some additional perspective on the significance of imaginary time:

Imaginary time in Hawking's view was the once and future time, and time as we know it but the broken-symmetry shadow of that original time. When a hand calculator cried "error" upon being asked the value of the square root of -1, it is telling us, in its way, that it belongs to this universe, and knows not how to inquire into the universe as it was prior to the moment of genesis. And that is the state of all science, until we have the tools in hand to explore the very different regime that pertained when time began.[44]

A Role for the Creator

Perhaps Hawking's most persistent concern is to analyze the role for a Creator in the universe that physics seems poised to define. In particular, he says that he, as Einstein before him, wants to know the mind of God. At one point he suggests that the success of scientific descriptions has led most people to believe that God allows the universe to evolve according to a set of laws and does not intervene to break those laws. This leaves the Creator with only the beginning of the universe as a province for activity. And yet at the point of a discussion of a no-boundary universe, he seems to see no role for God at all.[45]

At the conclusion to *A Brief History of Time,* Hawking continues

242 / Robert L. Herrmann

this line of reasoning, recalling that Einstein had once asked if God had had a choice in constructing the universe, and concluding that if the no-boundary proposal is correct, the answer would be no. Admittedly, God would still have had the freedom to choose the laws of the universe, although even here, it is suggested that God would be constrained by the paucity of theories that hold any promise to explain all the complicated structures of the universe, including ourselves. This seems an odd way to talk about God, as though his creativity depended on our ingenuity. But Hawking does go on to the larger questions of why we and the rest of the universe exist. The answer, he says, will be science's and human reason's ultimate triumph, for we shall know the mind of God.

For another perspective, the complementary view of origins that theology provides lends depth and breadth to any consideration of the mind of God. This universe is here for a purpose; it bears a grand design. And if we really wish to know God's thoughts, we must be prepared to be humbled by our own ignorance of the awesome immensity of what he has spoken into being.

Here, just as with our consideration of particles and fields, we are confronted by a universe that surprises us with its intricacy and its elusiveness. For James Maxwell, his brilliant conception of electromagnetism as a field of force was not a cause to set God aside. But rather his faith in God was the very basis for examining novel explanations for his data to see if they measured up to the richness to be expected from the Creator's mind.

Likewise, Schwinger cautions us to maintain a deep sense of humility in the present rush toward a unified field theory. The presumption of full knowledge, to assume that we have an adequate understanding of how nature works—what all the energies are—denies the entire history of physics. Perhaps some day we will have, as Hawking dearly hopes, a complete theory of the universe. But before it will gain the wider acceptance of the physics community, that theory will have to be backed by experimental data. Thus far, as Schwinger and Dyson have said, there is very little of the latter. And, as we have said, that data, as all of our scientific data, will lack the level of precision we once expected of it. In the final analysis, it is likely that even the most sophisticated and well-substantiated unified field theory will not give us more than a tiny inkling of the mind of God. For it will be our theory, and so of necessity only a human idea that refers to and seeks to express a profoundly deeper reality.

Self-Organization and a Creative Universe

A Creative Universe

If the unity of nature is indeed one of the indispensable presuppositions of mankind, then we can appreciate the search for principles that link all the forces of nature in a grand beginning. But there is much more to our universe than its grand beginning in a big bang, however that is theorized to have come about. For what follows the beginning is a most astounding story of creativity throughout time, from the void of space to galaxies, planets, crystals, life, and people. The agency of this transformation has been called self-organization, and its exquisite products strongly suggest that something quite beyond random forces is involved. Paul Davies tells us in *The Cosmic Blueprint* that for three centuries science has been ruled by the Newtonian and thermodynamic paradigms, which describe the universe as either an unchanging and sterile machine or as a process moving inexorably toward degeneration and decay. Now we are presented with a new paradigm, a creative universe, whose processes are collective, cooperative, and organizational.[46]

The great edifice of physics that was built up in the seventeenth century through the work of such pioneers as Galileo Galilei and Isaac Newton is commonly referred to as classical mechanics. The great triumph of Newton was to demonstrate that his laws of motion correctly described the shapes and periods of planetary orbits. In time it was assumed that if every particle of matter is subject to Newton's laws, with its motions determined by initial conditions and the forces acting upon it, then everything that happens in the universe is fixed in every detail. The universe is clockwork! Everything that ever happened, is happening now, or will happen in the future has been unalterably determined from the first instant of time. The future is fixed. This was the sweeping implication of Newtonian mechanics.

In such a situation, what is the effect upon time? Davies tells us that if the future is determined by the present, then the future is in a way already contained in the present.[47] The present state of the universe contains the information for the future, and, by inversion of the argument, for the past too. All of existence is thus frozen in a single moment of time. Past and future have no real meaning. Nothing actually happens! Nothing changes! Such was the world view of Newtonian physics. Yet we know from our own everyday experience that the world is changing, evolving. Past and future have real meaning. Things are happening. Time flows.

Ilya Prigogine and Isabelle Stengers, in their book *Order Out of Chaos,* tell us that by the middle of the nineteenth century scientists were introducing new concepts that concerned heat engines and energy conversion in a new science called thermodynamics.[48] Two laws were elaborated, one describing the conservation of energy and a second law that introduced the idea of disorder or entropy. The second law explained the frequently observed inefficiency associated with energy conversion. Some of the energy had been converted to an unusable form representing the increased molecular disorder of the system.

The implication of the second law was far-reaching; there was an irreversible direction to natural processes. Time had an arrow. By contrast with classical mechanics, where time is reversible and past and future are indistinguishable, in thermodynamics, time is irreversible and past and future are distinctly different. Thus, we understand the irreversibility of all natural phenomena as a consequence of the second law. Cooked eggs cannot be uncooked; we cannot reverse our own aging, and rivers do not flow uphill.

But this conclusion seemed to carry with it a very dire prediction. The universe is doomed. It is running down, inexorably approaching the eventual loss of all useful energy. This decay was said to have cosmic proportions because the sun and the rest of the stars are burning up their reserves of nuclear fuel and will finally grow dim and cold. This gloomy prognosis is called the "heat death" of the universe.

The Origin of Diversity
But there is another arrow of time that lies alongside the thermodynamic arrow but points in the opposite direction. Its origin is mysteri-

ous, but its existence is strongly substantiated by the plethora of complexity and diversity of form all around us. The universe is progressing to ever more developed and elaborate states of matter and energy. This arrow should surely be called the optimistic arrow, as opposed to the pessimistic arrow of the second law.

How can these two opposite tendencies in the universe, neither of which can be denied, be reconciled? The answer is being found in a poorly studied category of Earth's processes that are generally defined as nonlinear, far-from-equilibrium, and irreversible. Such processes are complex and irregular and difficult to study. The usual approach of modeling by approximation to regular systems is generally unsatisfactory, and the attitude on the part of most scientists has been to ignore complex reaction systems as unusual cases.

Davies tells us that in fact the majority of processes in nature are nonlinear and that we must seek new approaches for their study:

> There is a tendency to think of complexity in nature as a sort of annoying aberration which holds up the progress of science. Only very recently has an entirely new perspective emerged, according to which complexity and irregularity are seen as the norm and smooth curves the exception. In the traditional approach one regards complex systems as complicated collections of simple systems. That is, complex or irregular systems are in principle analyzable into their simple constituents, and the behavior of the whole is believed to be reducible to the behavior of the constituent parts. The new approach treats complex or irregular systems as primary in their own right. They simply cannot be "chopped up" into lots of simple bits and still retain their distinctive qualities.
>
> We might call this new approach synthetic or holistic, as opposed to analytic or reductionist, because it treats systems as wholes. Just as there are idealized simple systems (e.g., elementary particles) to use as building blocks in the reductionist approach, so one must also search for idealized complex or irregular systems to use in the holistic approach. Real systems can then be regarded as approximations to these idealized complex or irregular systems.

The new paradigm amounts to turning three hundred years of entrenched philosophy on its head. To use the words of physicist Predrag Cvitanovic, "Junk your old equations and look for guidance in clouds' repeating patterns." It is, in short, nothing less than a brand new start in the description of nature.[49]

The salient thing about these complex processes is that they can give rise to new structures and new order spontaneously. For example, the flow of water from a faucet goes through a series of ordered structures as the flow rate is changed. Superconductivity in metals at certain low temperatures reveals a collective ordering of electrons. And living systems reveal a high level of order in far-from-equilibrium conditions. Many of the complex reactions fit into the category of chaotic behavior, and it is fascinating to realize that order and chaos can be so closely related. One of the most fascinating visual effects is seen in a chemical phenomenon called the "Belousov-Zhabotinsky reaction." When Malonic acid, bromate, and cerium ions are placed in sulfuric acid in a shallow dish at certain critical temperatures, a series of pulsating concentric and spiral circles develop almost as if they were life forms. This behavior is the result of giant oscillations of millions of molecules, operating in concert, in the reaction system. The key feature of this kind of chemical phenomenon is the presence of autocatalysis: One or more of the reacting species is able to catalyze its own synthesis, and the whole system seems to pivot on this autocatalytic step.

In contrast to chemical reactions in which reagents and products are distributed randomly in the solution, in the Belousov-Zhabotinsky type of reaction there are local inhomogeneities; in one region, one component may predominate, while in another part of the reaction vessel its concentration may be exhausted. The cooperative effect of a vast number of such reactant molecules leads to the pulsating, highly ordered arrangements. Prigogine and Stengers propose that these structures are the inevitable consequences of far-from-equilibrium reactions. The term Prigogine uses for this phenomenon is order through fluctuation.[50]

What has been found is that the existence of complex and intricate structures does not require complicated and fundamental principles. Simple equations that can be handled by a pocket calculator can give rise to an extraordinarily rich variety and complexity. And these relationships are true for such diverse phenomena as turbulence in water and in the atmosphere, fluctuations of insect populations, behavior of a pendulum, and a chemical clock. Furthermore, such complex, dynamic systems very often display chaotic behavior. Finally, the evolution of such systems is exceedingly sensitive to the initial conditions, so that their behavior is essentially unpredictable.

This is not to imply that the behavior of chaotic systems is intrinsically nondeterministic. If we could fix the initial conditions, the entire future behavior of the system could be predicted precisely. But in practice we cannot specify the system exactly, and for chaotic systems this inaccuracy leads to a magnification of the error. In linear systems, errors grow in proportion to time. In chaotic systems, the errors grow at an escalating rate, exponentially with time.[51] But this essentially random element in complex systems is just the sort of situation one would desire for a truly creative mechanism, especially since the occurrence of chaos frequently goes hand in hand with the spontaneous generation of new spatial forms and structures. So nature can be both deterministic in principle and random. But for all practical purposes, determinism is a myth.

A Deepening Reality

We have seen that science has gone through a healthy reevaluation of its understanding of the nature of truth over the past two decades. No longer would most scientists claim that their theories represent anything more than approximations of the structures of reality. We have looked also at some of the current research in physics and chemistry and found that our goal of simplification and unification has been thwarted by drastic limitations in our capacity to provide experimental verification and by undreamt of levels of complexity in the systems under study.

In the physics of subatomic particles, we have achieved some simplification but at the same time we have discovered a strange new kind of complexity in a multiplicity of quarks and virtual particles. Our theoretical models for the force fields associated with these entities face severe limitations by virtue of calculations that attribute infinite mass and energy to electrons and because there is so little experimental verification.

We have seen a similar situation in the effort to achieve a unified theory of the forces of nature. The approach has been to examine the forces that were acting as time was moved back further and further in the direction of the moment of creation. It was postulated that the present universe may have arisen by a process of symmetry breaking, with the four forces of nature all together in a perfectly symmetrical relationship at the moment of origin of the universe. The present universe would then have evolved by a series of symmetry breakings, of which the break that produced the electromagnetic and weak nuclear forces was the most recent and least energetic.

The problem of further proof—the substantiation of so-called grand unified theories in which all four of the forces were one at the creation—has been addressed without success over the past two decades. The negative results from the very few experimental approaches to the problem have led many physicists to criticize the concept as too premature and too theoretical. Yet, its failure has not prevented physicist Hawking from claiming that the concept, in the particular refinement he calls quantum gravity, is the ultimate statement of physical reality. Again the mathematical formulation involves major complexities, of which the use of "imaginary time" is perhaps the most difficult conceptually.

All these efforts toward unification seem to many physicists and cosmologists to be extremely premature. Schwinger concludes that the rush for a unified theory represents an overt expression of hubris. We know so little about the energies involved and the way the universe works. We have been surprised again and again in the past by fundamentally new understandings of how nature works, and it would seem perilous to assume that symmetry breaking gives us anything even approaching a clear picture of reality.

Finally, we have looked at the new arrow of time, the thrust of a creative universe toward greater complexity and diversity. Again, the creation seems much more open-ended and innovative than we could have imagined, and its study seems a new beginning in the physics-chemical understanding of our universe.

The present situation in the sciences seems to shout for caution in our statements of what is, and how it came to be. What seems abundantly clear is that reality is much deeper and more profound then we had thought. Clerk Maxwell appears to have been "on the mark" when he suggested that appropriateness and tenability of scientific theories might best be evaluated by whether they measured up as far as possible to the "riches of creation." If we are attempting to think the Creator's thoughts, our thoughts must surely be only the simplest inklings of what really lies behind this vast universe and its awesome Creator.

Notes

1. John Templeton and Robert Herrmann, *Is God the Only Reality?* (New York: Continuum Publishing Group, 1994).

2. Timothy Ferris, *Coming of Age in the Milky Way* (New York: William Morrow, 1988) pp. 382-383.

3. Arthur Peacocke, *Intimations of Reality: Critical Realism in Science and Religion* (Notre Dame, Ind.: University of Notre Dame Press, 1984).

4. Mary Hesse, *Revolutions and Reconstructions in the Philosophy of Science* (Brighton: Harvester Press, 1981) p. xii

5. Thomas Kuhn, *Structures of Scientific Revolutions,* 2nd ed. (Chicago: University of Chicago Press) 1970.

6. Peacocoke, *Intimations of Reality,* p. 18.

7. Ernan McMullin, "The Relativist Critique of Science" in *The Sciences and Theology in the Twentieth Century,* ed. Arthur Peacocke (Notre Dame, Ind.: University of Notre Dame Press, 1981) pp. 301-2.

8. K.C. Cole, "On Imagining the Unseeable," *Discover,* December 1982, p.70.

9. Michael Polanyi, *Personal Knowledge: An Introduction to Post-Critical Philosophy* (New York: Harper & Row, 1966).

10. Walter Thorson, "The Biblical Insights of Michael Polanyi," *Journal of the American Scientific Affiliation,* 33(1981): 135.

11. Ernan McMullin, "A Case for Scientific Realism," in *Scientific Realism,* ed. Jarrett Leplin (Berkeley, Calif.: University of California Press, 1984) p. 26.

12. Peacocke, *Intimations of Reality,* pp. 26,27.

13. Ibid., p. 28.

14. Ian Hacking, *Representing and Intervening* (Cambridge, England: Cambridge University Press, 1987) p. 275.

15. Janet Soskice, *Metaphor and Religious Language* (Oxford, England: Oxford University Press, 1987) p. 115.

16. Peacocke, *Intimations of Reality,* pp.23,32.

17. Richard Morris, *The Nature of Reality* (New York: Noonday Press, Farrar, Straus and Giroux, 1987) p. 193.
18. Donald MacKay, *The Clockwork Image* (Downers Grove, Ill.: Intervarsity Press, 1974) pp. 44-45.
19. Arthur Peacocke, *God and the New Biology* (San Francisco: Harper & Row, 1986) pp. 21,22,25.
20. Robert Crease and Charles Mann, "How the Universe Works," *Atlantic Monthly*, August 1984, pp.68-69.
21. Morris, *Nature of Reality*, pp.74-78.
22. Ibid., p. 53.
23. Crease and Mann, "How the Universe Works," p. 70.
24. Morris, *Nature of Reality*, p. 80-81.
25. Ibid., p. 81.
26. Robert Crease and Charles Mann, *The Second Creation* (New York: Macmillan, 1986) p. 412.
27. Ferris, *Coming of Age*, pp.185-186.
28. Thomas Torrance, *A Dynamical Theory of the Electromagnetic Field* (Edinburgh: Scottish Academy Press, 1982) p. x.
29. Ferris, *Coming of Age*, pp.298-299.
30. Ibid., p. 299.
31. Stephen Hawking, *A Brief History of Time* (Toronto: Bantam, 1988) pp. 71-72.
32. Ibid., p. 74.
33. Crease and Mann, "How the Universe Works," p. 92.
34. Ferris, *Coming of Age*, p. 327.
35. Crease and Mann, *The Second Creation*, p. 411.
36. Ferris, *Coming of Age*, p. 388.
37. Hawking, *Brief History of Time*, pp.83-87.
38. Ibid., pp. 50-51.
39. Ibid., p. 133.
40. Ferris, *Coming of Age*, pp.363-364.
41. Hawking, *Brief History of Time*, pp. 137-38.
42. Ibid., p. 138.
43. Ibid., p. 139.
44. Ferris, *Coming of Age*, p. 364.
45. Hawking, *Brief History of Time*, pp. 140-141.
46. Paul Davis, *The Cosmic Blueprint* (New York: Simon and Shuster, 1988) p.2.
47. Ibid., p. 14.
48. Ilya Prigogine and Isabelle Stengers, *Order Out of Chaos* (New York: Bantam, 1984).
49. Davis, *Cosmic Blueprint*, pp. 22-23.
50. Prigogine and Stengers, *Order Out of Chaos*, p. 143.
51. Davis, *Cosmic Blueprint*, pp. 52-54.

John D. Barrow is professor of astronomy and director of the Astronomy Centre at the University of Sussex in England. He is the author of more than 230 research papers and a number of highly acclaimed books about the nature and significance of modern developments in physics and astronomy, including *The Left Hand of Creation, The Anthropic Cosmological Principle, The World Within the World, Theories of Everything: The Quest for Ultimate Explanation, The Origin of the Universe,* and *The Artful Universe.* Dr. Barrow's recent book, *Pi in the Sky,* explores the nature of mathematics as a human contrivance or as part of the mind of God.

Herbert Benson, M.D., is the chief of the division of behavioral medicine at Deaconess Beth Israel Medical Center, founding president of the Mind/Body Medical Institute, and the Mind/Body Medical Institute Associate Professor of Medicine, Harvard Medical School. Dr. Benson is the celebrated author of *The Relaxation Response, Beyond the Relaxation Response,* and with Marg Stark, *Timeless Healing: The Power and Biology of Belief.*

Freeman J. Dyson has been a professor of physics at the Institute for Advanced Study in Princeton, New Jersey, since 1953. Born in England, he first came to the United States as a Commonwealth Fellow at Cornell University. He is a Fellow of the Royal Society and a 1985 Gifford Lecturer. His writing in recent years has focused on issues of arms control and the ethical dilemmas of the nuclear age. Professor Dyson has also written about diversity in a 1988 book entitled *Infinite in all Directions,* for which he received the Phi Beta Kappa Award in Science.

Owen Gingerich is professor of astronomy and the history of science at Harvard University and a senior astronomer at the Smithsonian Astrophysical Observatory. In 1992–93, he chaired Harvard's History of Science Department. Professor Gingerich's research interests have ranged from the recomputation of an ancient Babylonian mathematical table to the interpretation of stellar spectra. He is a leading authority on Johannes Kepler and Nicholas Copernicus. Professor Gingerich has given the George Darwin Lecture and an Advent sermon at the National Cathedral in Washington, D.C. He has also witnessed eight total solar eclipses.

Robert L. Herrmann is adjunct professor of chemistry and chairman of the premedical program at Gordon College in Massachusetts and past executive director of the American Scientific Affiliation. He graduated from Purdue University and received his Ph.D. in biochemistry from Michigan State University in 1956. He taught medical school biochemistry for twenty-five years before beginning a career seeking to integrate religion and science. Dr. Herrmann has written two books with John Marks Templeton, *The God Who Would Be Known* and *Is God the Only Reality?*, and a chapter for another Templeton-edited book, *Looking Forward: The Next 40 Years.* He is a trustee of the John Templeton Foundation.

Martin E. Marty is Fairfax M. Cone Distinguished Service Professor of the History of Modern Christianity at the University of Chicago and the founder, past president, and senior scholar of the Park Ridge Center for Health, Faith and Ethics. He received his Ph.D. from the University of Chicago in 1956, and he taught at the University's Divinity School, the department of history, and in the program of the committee of the history of culture. Dr. Marty holds a divinity degree from the Lutheran School of Theology in Chicago. Internationally acclaimed for books on the history of religion and on fundamentalism, he edits *Context* newsletter.

Robert J. Russell is founder and director of the Center for Theology and the Natural Sciences and professor of theology and science in residence at the Graduate Theological Union in Berkeley, California. In 1968, he graduated from Stanford University with a major in physics and minors in music and religion. He then began concurrent studies in physics and theology, receiving an M.A. in theology from the Pacific School of Religion in Berkeley in 1972. His Ph.D. in experimental solid state physics was received from the University of California, Santa Cruz in 1978, on the same day that he was ordained in the United Church of Christ (Congregational). Dr. Russell has co-edited *Physics, Philosophy & Theology: A Common Quest for Understanding, Quantum Cosmology and the Laws of Nature: Scientific Perspective on Divine Action,* and *Chaos and Complexity: Scientific Perspectives on Divine Action.*

F. Russell Stannard is professor of physics and former head of department at the Open University, Milton Keynes, United Kingdom. London-born, he was educated at University College London, receiving his Ph.D. in cosmic ray physics in 1956. In 1983, he began studies of the relationship between science, religion, psychology, and philosophy. More recently, he has explored better ways to teach physics to young children and to incorporate modern thinking into school religious education lessons. Dr. Stannard's recent books in science and religion include *Grounds for Reasonable Belief* and a children's book, *Black Holes and Uncle Albert.* He is a trustee of the John Templeton Foundation.

Marg Stark is a freelance writer and author who lives in Silverdale, Washington. Her second book, *What No One Tells the Bride,* will be published in 1998. She is a frequent contributor to magazines, and her story about two men with AIDS was adapted into a 1994 NBC TV movie. Ms. Stark graduated from Mount Holyoke College and received her master's in journalism from Northwestern University. The daughter of a Presbyterian minister, she is especially interested in the power of faith to heal illness and divisions among people.

John Marks Templeton is chairman of the Templeton Foundations and president of the First Trust Bank, Nassau, Bahamas. He is the former president and director (1954-1975) of the Templeton Growth Fund Canada, Ltd. and Templeton World Fund, Inc. (1975-1995).

The donor of the Templeton Prizes for Progress in Religion, he is active in encouraging and financing the pursuit of integrative information in theology and the natural sciences and values in education. He has written, co-authored, and edited numerous books on science and religion.

Howard J. Van Till is professor of physics and astronomy in the physics department of Calvin College, Grand Rapids, Michigan. A graduate of Calvin College, he received his Ph.D. in physics from Michigan State University in 1965. His research has concerned both solid state physics and millimeter-wave astronomy, and he is a member of the American Astronomical Society. During the past decade, he has devoted considerable effort to writing on the relationship between natural science and Christian belief. He is the author of *The Fourth Day—What the Bible and the Heavens are Telling Us about the Creation,* and co-author of *Science Held Hostage* and *Portraits of Creation.* Dr. Van Till is a past president of the American Scientific Affiliation.

255